Principles of Plant Genetics and Breeding

Principles of Plant Genetics and Breeding

Nina Duran

R CALLISTO REFERENCE

www.callistoreference.com

Callisto Reference,
118-35 Queens Blvd., Suite 400,
Forest Hills, NY 11375, USA

Visit us on the World Wide Web at:
www.callistoreference.com

ISBN: 978-1-64116-229-6 (Hardback)

Cataloging-in-Publication Data

Principles of plant genetics and breeding / Nina Duran.
    p. cm.
Includes bibliographical references and index.
ISBN 978-1-64116-229-6
1. Plant genetics. 2. Plant breeding. 3. Genetics. 4. Breeding. I. Duran, Nina.
QK981 .P75 2019
581.35--dc23

Table of Contents

	Preface	**VII**
Chapter 1	**Introduction to Plant Breeding**	**1**
	• Plant Genetics	1
	• Plant Breeding	10
Chapter 2	**Plant Development and Growth**	**14**
	• Plant Development	14
	• Plant Morphology	21
	• Plant Embryogenesis	29
	• Plant Sporogenesis	35
	• Epigenetics of Plant Growth and Development	41
	• Plant Photomorphogenesis	45
	• ABC Model of Flower Development	55
Chapter 3	**Reproductive Systems**	**60**
	• Plant Reproduction	60
	• Vegetative Propagation	67
	• Asexual Reproduction	69
	• Sexual Reproduction	81
Chapter 4	**Crop Improvement: Genetic Concepts and Principles**	**98**
	• Mitosis and Meiosis	98
	• Plant Chromosome	104
	• Linkage	109
	• Chromosomal Crossover	113
	• Mutations	119
	• Gene Interaction	124
	• Pleiotropy	132
	• Penetrance and Expressivity	132
	• Cytoplasmic Inheritance	133
	• Hybridization	139

Chapter 5   **Selection Methods and Breeding**   145

      • Breeding of Self-pollinated Plants   145

      • Breeding of Cross-pollinated Plants   167

      • Breeding of Hybrid Cultivars   188

      • Breeding of Clonally Propagated Plants   202

Permissions

Index

Preface

The study of genes, variation and heredity in plants is under the scope of plant genetics. An important area of study in plant genetics is plant breeding. It is the practice of altering or enhancing certain traits in plants to obtain desired characteristics. Some of these include disease resistance, higher yield, drought tolerance or better adaptability to changed climatic condition. Modern plant breeding employs techniques such as marker assisted selection, reverse breeding and double haploids. Genetic modification that allows the addition or deletion of new genes to produce desirable phenotypes in plants is another method used for plant breeding. An understanding of plant genetics and plant breeding techniques can enable the development of solutions for the sustainment of agriculture in the face of harsh cropping conditions, food security concerns or loss of soil quality. This book unfolds the innovative aspects of plant breeding which will be crucial for the holistic understanding of the subject matter. It further elucidates the concepts and innovative models around prospective developments with respect to plant genetics. Coherent flow of topics, student-friendly language and extensive use of examples make this book an invaluable source of knowledge.

To facilitate a deeper understanding of the contents of this book a short introduction of every chapter is written below:

Chapter 1- The study of plant genes, variation and heredity is under the domain of plant genetics. It is studied under the field of botany. This chapter closely examines the key concepts of plant genetics and plant breeding, to provide an extensive understanding of the subject.

Chapter 2- A plant grows new tissues and structures from meristems, which are located between mature tissues or at the tips of organs. The study of plant growth and development requires an understanding of plant morphology, embryogenesis, sporogenesis, and photomorphogenesis, which have been covered in extensive detail in this chapter.

Chapter 3- In flowering plants, the flower is the primary reproductive structure in a plant. It shows great diversity in form and structure and also in methods of reproduction. In flowering and non-flowering plants, there exists a complex interplay between morphological adaptation and influence of environmental factors in sexual reproduction. This chapter has been carefully written to provide an extensive understanding of the varied aspects of sexual and asexual reproduction, and vegetative reproduction in plants.

Chapter 4- In agriculture, crops are modified using genetic engineering. A new trait is introduced or some existing trait is improved for better agricultural productivity. The topics elaborated in this chapter such as mitosis, meiosis, structural chromosomal, chromosome number, chromosomal crossover, gene interaction, cytoplasmic inheritance, pleiotropy, mutations, etc. will help in providing a better understanding of the role of genetics for enhanced crop production.

Chapter 5- The mode of reproduction in a plant is responsible for its genetic composition.

This in turn determines the suitable breeding or selection method in the plant. The aim of this chapter is to explore the different selection and breeding methods of self-pollinated, cross-pollinated, hybrid and clonally propagated plants.

I owe the completion of this book to the never-ending support of my family, who supported me throughout the project.

Nina Duran

Chapter 1

Introduction to Plant Breeding

The study of plant genes, variation and heredity is under the domain of plant genetics. It is studied under the field of botany. This chapter closely examines the key concepts of plant genetics and plant breeding, to provide an extensive understanding of the subject.

Plant Genetics

Plant genetics is a branch of genetics that studies heredity and variation in the higher plants. The following methods are used to study plant genetics, in addition to those used in other branches of genetics (hybridological analysis in particular). First, the role of each chromosome in the inheritance and development of differentplant characters is determined by monosomic analysis. This method, worked out on the thorn apple, is used to study several allopolyploids (such as certain wheats and cotton) as well as diploids (barley). Experimental mutagenesis is very valuable inplant genetics because it produces a large variety of new forms used in breeding, as well as material useful for studying thegenetics of individual plant species. Genetic chromosome maps are constructed with the help of mutants. With these mapsone can investigate the effect of an altered gene (in a homozygous or heterozygous state) on the development of individualcharacters under various environmental conditions and on the physiological and biochemical properties of plants. The study of mutants helps one to clarify the evolution of a particular species.

Two methods of investigating the evolution of piants are hybridization and analysis of chromosome conjugation in hybridsduring meiosis. (Unrelated chromosomes do not conjugate.) An important method is the artificial resynthesis of existingspecies by hybridization and the subsequent doubling of the number of chromosomes. Allopolyploidy plays an important rolein the evolution of plants, including many cultivated ones such as wheat, oats, cotton, potatoes, and fruits. After it wasdiscovered that the alkaloid colchicine can prevent doubled chromosomes from going to different poles of the cell, autopolyploidy was widely used to obtain new and sometimes very valuable forms. Distant hybridization combined withcytogenetics is used in the study of the role of individual chromosomes and segments of chromosomes in the inheritance ofcharacters. It is also used to devise techniques for inserting the segments of chromosomes of wild plants responsible for thedevelopment of valuable characters, such as resistance to rust, into the chromosomes of cultivated plants. The role of thenucleus and the cytoplasm in the inheritance and development of characters is studied by means of distant hybridization andan analysis of the nature of male cytoplasmic sterility used to obtain heterotic forms. Plant geneticists investigate apomixisand self-incompatibility, that is, the incapacity of plants for self-fertilization, as well as the genetic peculiarities of self and cross-pollinating plants and vegetatively and apomictically reproducing forms.

Plant genetics is being increasingly influenced by the ideas and methods of molecular biology (such as the hybridization of DNA and DNA—RNA and the study of isoenzymes). The methods of population genetics and biometrics are used in plant genetics to differentiate the genotypic and paratypic elements in the general phenotypic variation of characters, thus increasing the effectiveness of artificial selection. All these methods are used to improve the economically valuable properties of crops: yielding capacity, resistance to unfavorable environmental conditions, some biochemical and-technological characteristics of a plant or its grain, and developmental characteristics (its condition in winter and in spring, early ripening, and so forth). Among the higher plants that are studied most from the genetic standpoint are corn, mouse-earcress (a plant of the family Cruciferae known as the plant drosophila, a model object of genetic research), peas, tomatoes, and barley. Hybridization has been used in these plants to determine the location of genes and to compile chromosomemaps. Under intensive study is the cytogenetics of bread wheat, a complex 42-chromosome allopolyploid that evolved in thecourse of the natural hybridization of three different cereal grasses, with subsequent doubling of the chromosome number inhybrids. Plant genetics has contributed greatly to breeding. For example, it includes the use of heterosis in the breeding ofcorn based on male sterility, the introduction of genes responsible for the high lysine content of the grain into high-yieldinghybrids and fodder barley varieties, the creation of low, nonlodging, high-yielding wheat varieties by means of dwarfismgenes (the "green revolution" in India and other countries), and the breeding of productive and sacchariferous triploid sugarbeet hybrids.

Plant Genomes

Plant genomes are best described in terms of genome size, gene content, extent of repetitive sequences and polyploidy/duplication events. Although plants also possess mitochondrial and chloroplast genomes, their nuclear genome is the largest and most complex. There is extensive variation in nuclear genome size (Table) without obvious functional significance of such variation.

Nuclear genome size in different species

Common name	Scientific name	Nuclear genome size
Wheat	Triticum aestivum	15,966
Onion	Allium cepa	15,290

Garden pea	Pisum sativum	3,947
Corn	Zea mays	2,292
A sparragus	As paragus officinalis	1,308
Tomato	Lycopersicum esculentum	907
Sugarbeet	Beta vulgaris	758
Apple	Malus X domestica	743
Common bean	Phaseolus vulgaris	637
Cantal oupe	Cucumis melo	454
Grape	Vitis vinifera	483
Man	Homo sapiens	2,910

Expressed in Megabases (1 Mb: 1,000,000 bases)

Plant genomes contain various repetitive sequences and retrovirus-like retro trans-posons containing long terminal repeats and other retro elements, such as long in-terspersed nuclear elements and short-interspersed nuclear elements. Retro element insertions contribute to the large difference in size between collinear genome segments in different plant species and to the 50% or more difference in total genome size among species with relatively large genomes, such as corn. They contribute a smaller percent-age of genome size in plants with smaller genomes such as *Arabidopsis*. If other repet-itive sequences are accounted for, the corn genome is comprised of over 70% repetitive sequences and of 5% protein encoding regions.

It is widely accepted that 70-80% of flowering plants are the product of at least one polyploidization event. Many economically important plant species, such as corn, wheat, potato, and oat are either ancient or more recent polyploids, comprising more than one, and in cases such as wheat, three different homologous genomes within a single species. Duplicated segments also account for a significant fraction of the rice genome. About 60% of the *Arabidopsis* genome is present in 24 duplicated segments, each more than 100 kilo bases (kb) in size. Ancestral polyploidy contributes to create genetic variation through gene duplication and gene silencing. Genome duplication and subsequent divergence is an important generator of protein diversity in plants.

Model Plant Species

Model organisms (*Drosophila melanogaster, Caenorhabditis elegans, Saccharomyces cerevisiae*) provide genetic and molecular insights into the biology of more complex species. Since the genomes of most plant species are either too large or too complex to be fully analyzed, the plant scientific community has adopted model organisms. They share features such as being diploid and appropriate for genetic analysis, being amenable to genetic transformation, having a (relatively) small genome and a short growth cycle, having commonly available tools and resources, and being the focus of research by a large scientific community. Although the advent of tissue culture techniques fostered the use of tobacco and petunia, the species now used as model organisms for mono-and dicotyledonous plants are rice (*Oryza sativa*) and *Arabidopsis* (*Arabidopsis thaliana*) respectively.

Arabidopsis, a small Cruciferae plant without agricultural use, sets seed in only 6 weeks from planting, has a small genome of 120 Megabases (Mb) and only five chromosomes. There are extensive tools available for its genomic analysis, whole genome sequence, Expressed Sequence Tags (ESTs) collections, characterized mutants and large populations mutagenized with insertion elements (transposons or the T-DNA of *Agrobacterium*). *Arabidopsis* can be genetically transformed on a large scale with *Agrobacterium tumefaciens* and biolistics. Other tools available for this model plant are saturated genetic and physical maps.

Unlike *Arabidopsis*, rice is one of the world's most important cereals. More than 500 million tons of rice is produced each year, and it is the staple food for more than half of the worldís population. There are two main rice subspecies. *Japonica* is mostly grown in Japan, while *indica* is grown in China and other Asia-Pacific regions. Rice also has very saturated genetic maps, physical maps, whole genome sequences, as well as EST collections pooled from different tissues and developmental stages. It has 12 chromosomes, a genome size of 420 Mb, and like *Arabidopsis*, it can be transformed through biolistics and *A. tumefaciens*. Efficient transposon-tagging systems for gene knockouts and gene detection have not yet become available for saturation mutagenesis in rice, although some recent successes have been reported.

Maps

Genetic Maps

The development of molecular markers has allowed for constructing complete genetic maps for most economically important plant species. They detect genetic variation directly at the DNA level. A myriad of molecular marker systems are available, yet their description lies beyond the scope of this paper. A genetic map represents the ordering of molecular markers along chromosomes as well as the genetic distances, generally expressed as centiMorgans (cM), existing between adjacent molecular markers. Genetic maps in plants have been created from many experimental populations, but the most frequently used are F2, backcrosses and recombinant inbred lines. Although longer to develop, recombinant inbred lines offer a higher genetic resolution and practical advantages. Once a mapping population has been created, it takes only few months to produce a genetic map with a 10 cM resolution. Genetic maps contribute to the understanding of how plant genomes are organized and once available they facilitate the development of practical applications in plant breeding, such as the identification of

Quantitative Trait Loci and Marker Assisted Selection. Most economically important plant traits such as yield; plant height and quality components exhibit a continuous distribution rather than discrete classes and are regarded as quantitative traits. These traits are controlled by several loci each of small effect and different combinations of alleles at these loci can give different phenotypes.

Quantitative Trait Loci analysis refers to the identification of genomic regions associated with the phenotypic expression of a given trait. Once the location of such genomic regions is known they can be assembled into designer genotypes, i.e. individuals carrying chromosomic fragments associated with the expression of a given phenotype. The most important feature of Marker Assisted Selection is that once a molecular marker genetically linked to the expression of a phenotypically interesting allele has been detected, an indirect selection for such allele based upon the detection of the molecular marker can be accomplished, since little or any genetic recombination will occur between them. Therefore, the presence of the molecular marker will always be associated with the presence of the allele of interest.

Genetic maps are also an important resource for plant gene isolation, as once the genetic position of any mutation is established, it is possible to attempt its isolation through positional cloning. Further more, genetic maps help establish the extent of genome colinearity and duplication between different species.

Physical Maps

Although genetic maps provide much-needed landmarks along chromosomes, they are still too far apart to provide an entry point into genes, since even in model plants the kilo bases per centiMorgan (kb/cM) ratio is large, from 120 to 250 kb/cM in *Arabidopsis* and between 500 and 1.500 kb/cM in corn. Therefore, a 1 cM interval may harbor ~30 to 100 or even more genes. Physical maps bridge such gaps, representing the entire DNA fragment spanning the genetic location of adjacent molecular markers.

Physical maps can be defined as a set of large insert clones with minimum overlap encompassing a given chromosome. First generation physical maps in plants were based on YACs (Yeast Artificial Chromosomes). Chimaerism and stability issues, however, dictated the development of low copy, *E. coli*-maintained vectors such as Bacterial Artificial Chromosomes (BACs) and P1-derived artificial chromosomes. Although BAC vectors are relatively small (molecular weight of BAC vector pBeloBAC11 is 7.4 kb for instance), they carry inserts between 80 and 200 kb on average and possess traditional plasmid selection features such as an antibiotic resistance gene and a polycloning site within a reporter gene allowing insertional inactivation. BAC clones are easier to manipulate than yeast-based clones. Once a BAC library is prepared, clones are assembled into contigs using fluorescent DNA fingerprint technologies and matching probabilities. Physical and genetic maps can be aligned, bringing along continuity from phenotype to genotype. Furthermore, they provide the platform clone-by-clone sequencing approaches rely upon. Figure shows the relationship between genetic and physical maps and their alignment. Physical maps provide the bridge needed between the resolution achieved by genetic maps and that needed to isolate genes through positional cloning.

Genome Colinearity/Genome Evolution

A remarkable feature of plant genomics is its ability to bring together more than one species for analysis. The comparative genome mapping of related plant species has shown that the organization of genes is highly conserved during the evolution of members of taxonomic families. This has led to the identification of genome colinearity between the well-sequenced model crops and their related species (e.g. *Arabidopsis* for dicots and rice for monocots). Colinearity overrides the differences in chromosome number and genome size and can be defined as conservation of gene order within a chromosomal segment between different species. A related concept is synteny, which refers to the presence of two or more loci on the same chromosome regardless they are genetically linked or not.

Colinear relationships have been observed among cereal species (corn, wheat, rice, barley), legumes (beans, peas and soybeans), pines and *Cruciferae* species (canola, broccoli, cabbage, *Arabidopsis thaliana*). Recently, the first studies at the gene level have demonstrated that microcolinearity of genes is less conserved; small-scale rearrangements and deletions complicate microcolinearity between closely-related species. For instance, although a 78-kb genomic sequence of sorghum around the locus *adh1* and its homologous genomic fragment from maize showed considerable microcolinearity and the fact that they share nine genes in perfect order and transcriptional direction, five additional, unshared genes reside in this genomic region.

Comparing sequences of soybean and *Arabidopsis* demonstrated partial homology between two soybean chromosomes and a 25 cM section of chromosome 2 from *Arabidopsis*. Although such relationships need to be assessed on a case-by-case basis, they reflect the value *Arabidopsis* and other model species offer to economically important species.

Colinearity has also been established between rice and most cereal species, allowing the use of rice for genetic analysis and gene discovery in genetically more complex species, such as wheat and barley. A comparison of rice and barley DNA sequences from syntenic regions between barley chromosome 5H and rice chromosome 3 revealed the presence of four conserved regions, containing four predicted genes. General gene structure was largely conserved between rice and barley. A similar comparison between corn and rice, based on 340 kb around loci *adh1* and *adh2*, showed five colinear genes between the two species, as well as a possible translocation on *adh1*. Rice genes similar to known disease resistant genes showed no cross-hybridization with corn genomic DNA, suggesting sequence divergence or their absence in maize. There are even reports of colinearity across the mono-dicotyledoneous division involving *Arabidopsis* and cereals which diverged as far back as 200 million years ago Exploiting colinearity helps to establish cross-species genetic links and also aids in the extrapolation of information from species with simpler genomes (i.e. rice) to genetically complex species (corn, wheat). Furthermore, it reflects the power of genomics to integrate genetic information across species.

Whole Genome Sequencing

Genetic and physical maps at the inter- or intra-species level represent a key layer of genomic information. However, sequence data represents the ultimate level of genetic information. Three major breakthroughs have allowed the sequencing of complete genomes:

- The development of fluorescence-based DNA sequencing methods that provide at least 500 bases per read;

- The automation of several processes such as picking and arraying bacterial subclones, purification of DNA from individual subclones and sample loading among others; and

- The development of software and hardware able to handle massive amounts (gigabytes) of data points.

There are two main approaches to large scale sequencing. In clone-by-clone strategies, large insert libraries, such as those based on BAC clones, are used as sequencing templates, and inserts are arranged into contigs using diverse fingerprinting methods to establish minimal tiling paths. Sequence Tagged Connectors extracted from large insert clones as well as FISH (Fluorescent *in situ* Hybridization) and optical mapping are used to extend contigs and close gaps. BAC clones from sequence-ready contigs are then fragmented into plasmid or M13 vector-based shotgun libraries with insert sizes of ~1 to 3 kb. Using more than one vector system reduces cloning bias issues. Sequencing efforts are tailored to the degree of coverage required. For instance, for a 5-fold coverage, and assuming 500 base pairs (bp) per sequencer reading, 800 clones are sequenced to cover an 80 kb BAC clone. Finished sequences are those obtained at a ~8-10 fold coverage and provide >99.99% accuracy, whereas working draft sequences are attained at a ~3-5 fold coverage. It is important to note, however, that even working draft sequences provide an enormous amount of information, and even shotgun approaches rely to some extent on clone-by clone information.

After sequencing is concluded, DNA data is used to reassemble BAC clones. Base calling programs assigning quality scores to each read base such as Phred, sequence assembly programs such as Phrap, and graphical viewing tools are used to achieve such assembly. The finishing of the sequence then ensues, which can be done in part manually or with finishing software such as Autofinish.

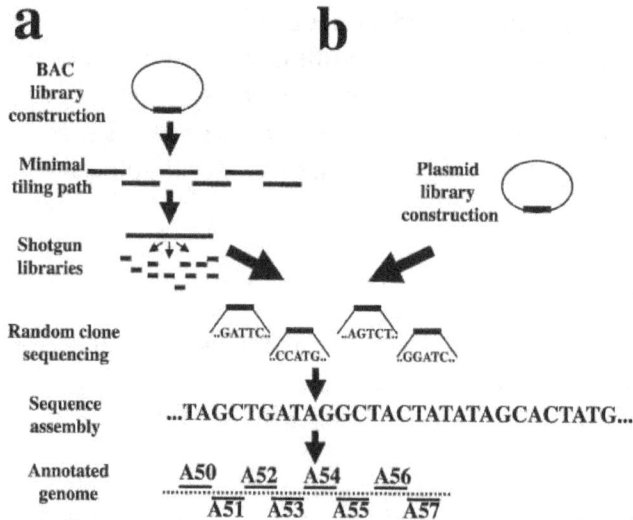

Figure: Whole genome sequencing a: Clone-by-clone approach; b: Shot-gun approach

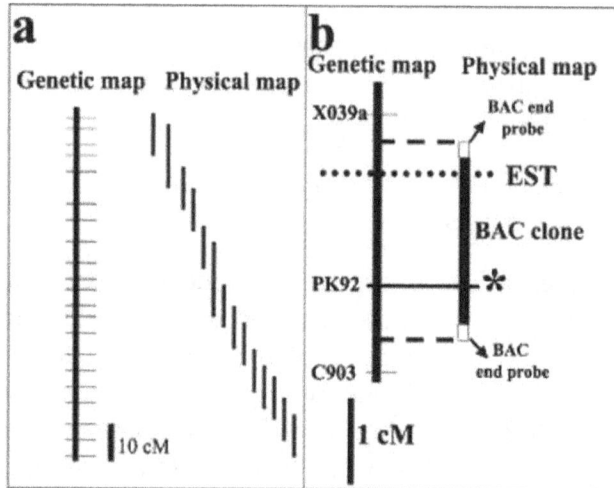

Figure: Maps used in plant genetics. a: Genetic and physical maps of a hypothetical chromosome. Horizontal lines on the genetic map represents loci targeted by a molecular marker; vertical lines represent overlapping BAC clones. b: Alignment of genetic and physical maps using BAC ends sequence (dashed lines), ESTs (dotted line) and molecular markers (*).

Annotation, or the process of identifying start and stop codons and the position of introns that permits the prediction of biological function from DNA sequence, proceeds through three main steps. The first is to use gene finders like Xgrail or others based on generalized hidden Markov models, such as GeneMark.hmm and Gen Scan, specifically developed to recognize *Arabidopsis* genes. In the second step, sequences are aligned to protein and EST databases; and finally, putative functions are assigned to each gene sequence. Successful annotation processes often combine different software and manual inspection.

In shotgun approaches, which have been successfully used to sequence many microorganisms and *D. melanogaster*, small insert libraries are prepared, and randomly selected inserts are sequenced until a ~5-fold or higher coverage is reached. Sequences are then assembled, gaps identified and closed, and finally annotation conducted. Shotgun sequencing does not rely upon the availability of minimal tiling paths and therefore reduces the cost and effort required to obtain whole genome sequences. Nevertheless, they require an enormous amount of computational power to assembly a large number of random sequences into a small number of contigs. Furthermore, the ultimate quality of large genomes that have been shotgun-sequenced may not be as high as that achievable using the clone-by-clone approach. Because of a high content of long and highly conserved repetitive sequences, including retro transposons, shotgun sequencing of plant genomes may pose special challenges.

The *Arabidopsis* genome was the first to be fully sequenced. The ecotype chosen was Columbia. In 1996, sequencing groups in the US, Japan and Europe established the *Arabidopsis* Genome Initiative (AGI) and set common techniques and resources, accuracy standards, levels of analysis, and a common public release policy for sequence information. Since shotgun sequencing was not available at its inception, the sequencing of *Arabidopsis* followed a more conventional approach.

There are two remarkable lessons to be learned from the *Arabidopsis* sequencing effort. First, there is no alternative to establishing partnerships between even competing groups when tackling large genomic projects because of the complexity, expense, and infrastructure required. For example, the effectiveness of the AGI consortium resulted in the completion and release of the *Arabidopsis* sequence in 2000, fully four years ahead of schedule. Secondly, approximately one third of the genes putatively identified in *Arabidopsis* encode products lacking significant similarity to proteins of known function in other organisms. Moreover, only 9% of its genes have been characterized experimentally. Such figures reflect the power of genomic approaches and the wealth of information they provide us with. The gene complement of *Arabidopsis* is shown in table.

In rice, the IRGSP (International Rice Genome Sequencing Project) started in 1997, and includes members from developing countries in addition to European and USA partners. It is based on the Nipponbare cultivar, and its approach is similar to that used in *Arabidopsis*. Thus, the first task was to establish a sequence-ready BAC contig of the rice genome, followed by software assembly of DNA sequences, computational and manual annotation and final release of the terminated sequence. The expected deadline for release of the full sequence data is 2003.

Syngenta and a Chinese group recently made available the sequences of japonica and *indica* rice, respectively, and both were based on shotgun approaches. Gradients in Guanine/Cytosine content and codon usage for rice genes created unexpected problems in the gene annotation process, and the gene finder software F gene SH was the most effective in rice. The number of genes predicted in rice ranges from 32,000 to 55,000, depending on the criteria used to recognize a gene. Regardless of the actual figure, it is interesting to note that such figures are similar or larger than the human gene complement (32,000-39,100 genes. The suggestion of that protein diversity in plants is generated primarily through gene duplication would account for the comparatively large number of genes predicted in rice.

Nevertheless, the actual number of genes existing in *Arabidopsis*, rice, or any other sequenced species remains to be established through functional genomic experiments that establish the biological meaning of DNA sequences, since gene prediction through homology comparisons and software tools is a statistical "best informed guess" rather than a biologically based process. Availability of extensive EST collections, which now exist in several plant species, including corn and soybean, reduce the dependence of the annotation process on computational gene predictions.

Table: Gene complement of Arabidopsis thaliana

Feature	Chromo-some 1	Chromo-some 2	Chromo-some 3	Chromo-some 4	Chromo-some 5
DNA molecule					
Length (dp)	29,105	19,646,945	23,172,617	17,549,867	25,953,409
Number of genes	6,543	4,036	5,220	3,825	5,874
Gene density (kb/gene)	4.0	4.9	4.5	4.6	4.4
Exons					
Average number per gene	5.4	4.9	5.1	5.2	5.3
Average size (dp)	247	259	250	256	242
Introns					
Average size (dp)	168	177	159	186	159
ESTs					
Proportion of genes with ESTs	60.8%	56.9%	59.8%	61.4%	61.4%
Number of EST	30,522	14,989	20,732	16,605	22,885

Plant Breeding

Plant breeding is the application of genetic principles to produce plants that are more useful to humans. This is accomplished by selecting plants found to be economically or aesthetically desirable, first by controlling the mating of selected individuals, and then by selecting certain individuals among the progeny. Such processes, repeated over many generations, can change the hereditary makeup and value of a plant population far beyond the natural limits of previously existing populations.

Plant breeding is an ancient activity, dating to the very beginnings of agriculture. Probably soon after the earliest domestications of cereal grains, humans began to recognize degrees of excellence among the plants in their fields and saved seed from the best for planting new crops. Such tentative selective methods were the forerunners of early plant-breeding procedures.

The results of early plant-breeding procedures were conspicuous. Most present-day varieties are so modified from their wild progenitors that they are unable to survive in nature. Indeed, in some cases, the cultivated forms are so strikingly different from existing wild relatives that it is difficult even to identify their ancestors. These remarkable transformations were accomplished by early plant breeders in a very short time from an evolutionary point of view, and the rate of change was probably greater than for any other evolutionary event.

Scientific plant breeding dates back hardly more than 50 years. The role of pollination and fertilization in the process of reproduction was not widely appreciated even 100 years ago, and it was not until the early part of the 20th century that the laws of genetic inheritance were recognized and a beginning was made toward applying them to the improvement of plants. One of the major facts that has emerged during the short history of scientific breeding is that an enormous wealth of genetic variability exists in the plants of the world and that only a start has been made in tapping its potential.

Nature of Plant Breeding

Plant breeding is the art and science of changing and improving the heredity of plants. In earlier days the extent of plant breeding as an art and as a science was much disputed Early Ln was a nomad and dependent for his food on the forest products.

As civilization progressed he learned to cultivate more plants and selected the seeds from healthier plants for sowing the next year. His method of selection was designed without an understanding of the principal of inheritance. Therefore, plant breeding then was largely an art (this selection is the earliest method of plant breeding and is practiced on a large scale, even today).

As man's knowledge about plants increased, he was able to select more intelligently with the discovery of sex in plants, knowledge about inheritance of characters role of environment in producing characters, the basis of variations in various plant characters addition of hybridization etc., plant breeding methods are based on scientific principles of plant science particularly of genetics and cytogenetics.

As the breeder's pledge of genetics and related plant sciences progressed, plant breeding became less of an art and more of a science. The modern plant breeding is, therefore, considered as science based upon a thorough understanding and use of genetic principles.

Goals of Plant Breeding

The plant breeder usually has in mind an ideal plant that combines a maximum number of desirable characteristics. These characteristics may include resistance to diseases and insects; tolerance to heat and frost; appropriate size, shape, and time to maturity; and many other general and specific traits that contribute to improved adaptation to the environment, ease in growing and handling, greater yield, and better quality. The breeder of fancy show plants must also consider aesthetic appeal. Thus the breeder can rarely focus attention on any one characteristic but must take into account the manifold traits that make the plant more useful in fulfilling the purpose for which it is grown.

Increase of Yield

One of the aims of virtually every breeding project is to increase yield. This can often be brought about by selecting obvious morphological variants. One example is the selection of dwarf, early maturing varieties of rice. These dwarf varieties are sturdy and give a greater yield of grain. Furthermore, their early maturity frees the land quickly, often allowing an additional planting of rice or other crop the same year.

Another way of increasing yield is to develop varieties resistant to diseases and insects. In many cases the development of resistant varieties has been the only practical method of pest control. Perhaps the most important feature of resistant varieties is the stabilizing effect they have on production and hence on steady food supplies. Varieties tolerant to drought, heat, or cold provide the same benefit.

Modifications of Range and Constitution

Another common goal of plant breeding is to extend the area of production of a crop species. A good example is the modification of grain sorghum since its introduction to the United States about 100 years ago. Of tropical origin, grain sorghum was originally confined to the southern Plains area and the Southwest, but earlier maturing varieties were developed until grain sorghum is now an important crop as far north as North Dakota.

Development of crop varieties suitable for mechanized agriculture has become a major goal of plant breeding in recent years. Uniformity of plant characters is very important in mechanized agriculture because field operations are much easier when the individuals of a variety are similar in time of germination, growth rate, size of fruit, and so on. Uniformity in maturity is, of course, essential when crops such as tomatoes and peas are harvested mechanically.

The nutritional quality of plants can be greatly improved by breeding. For example, it is possible to breed varieties of corn (maize) much higher in lysine than previously existing varieties. Breeding high-lysine maize varieties for those areas of the world where maize is the major source of this nutritionally essential amino acid has become a major goal in plant breeding.

In breeding ornamentals, attention is paid to such factors as longer blooming periods, improved keeping qualities of flowers, general thriftiness, and other features that contribute to usefulness and aesthetic appeal. Novelty itself is often a virtue in ornamentals, and the spectacular, even the bizarre, is often sought.

Evaluation of Plants

The appraisal of the value of plants so that the breeder can decide which individuals should be discarded and which allowed to produce the next generation is a much more difficult task with some traits than with others.

Qualitative Characters

The easiest characters, or traits, to deal with are those involving discontinuous, or qualitative, differences that are governed by one or a few major genes. Many such inherited differences exist, and they frequently have profound effects on plant value and utilization. Examples are starchy versus sugary kernels (characteristic of field and sweet corn, respectively) and determinant versus in determinant habit of growth in green beans (determinant varieties are adapted to mechanical harvesting). Such differences can be seen easily and evaluated quickly, and the expression of the traits remains the same regardless of the environment in which the plant grows. Traits of this type are termed highly heritable.

Quantitative Characters

In other cases, however, plant traits grade gradually from one extreme to another in a continuous series, and classification into discrete classes is not possible. Such variability is termed quantitative. Many traits of economic importance are of this type; e.g., height, cold and drought tolerance, time to maturity, and, in particular, yield. These traits are governed by many genes, each having a small effect. Although the distinction between the two types of traits is not absolute, it is nevertheless convenient to designate qualitative characters as those involving discrete differences and quantitative characters as those involving a graded series.

Quantitative characters are much more difficult for the breeder to control, for three main reasons:

- The sheer numbers of the genes involved make hereditary change slow and difficult to assess;

- The variations of the traits involved are generally detectable only through measurement and exacting statistical analyses; and

- Most of the variations are due to the environment rather than to genetic endowment; for example, the heritability of certain traits is less than 5 percent, meaning that 5 percent of the observed variation is caused by genes and 95 percent is caused by environmental influences.

Chapter 2

Plant Development and Growth

A plant grows new tissues and structures from meristems, which are located between mature tissues or at the tips of organs. The study of plant growth and development requires an understanding of plant morphology, embryogenesis, sporogenesis, and photomorphogenesis, which have been covered in extensive detail in this chapter.

Plant Development

Plant development is an umbrella term for a broad spectrum of processes that include: the formation of a complete embryo from a zygote; seed germination; the elaboration of a mature vegetative plant from the embryo; the formation of flowers, fruits, and seeds; and many of the plant's responses to its environment. Plant development encompasses the growth and differentiation of cells, tissues, organs, and organ systems. Plant development shares many similarities with developmental processes in animals, but the fact that plants are nonmotile, photosynthetic organisms requires certain novel developmental processes in addition to the common ones.

Embryo and Seed Development

Embryogenesis, the formation of a multicellular embryo from a single-celled zygote, is one of the most dramatic and best-characterized aspects of plant development. Four key developmental processes take place during embryogenesis. First, the zygote expresses apical -basal polarity, meaning that the apical and basal ends of the zygote cell differ structurally and biochemically. When the zygote divides, it typically divides asymmetrically, giving rise to a small apical cell with dense cytoplasm and a large basal cell with watery cytoplasm. Although these two cells have identical nuclei, their fates differ dramatically. The apical cell gives rise to the embryo itself, while the basal cell gives rise to a short-lived structure called a suspensor and the tip of the root system. The progeny of the apical cell grow and divide to form a spherical mass of cells, the globular-stage embryo.

Second, differential growth within the globular embryo gives rise to the "heart" stage embryo, the earliest stage when the precursors of cotyledons, root, and stem can be recognized. This key embryogenic process is called organogenesis. Third, distinctive planes of cell divisions bring about histogenesis, the process by which cells within embryonic cotyledons, root, and stem acquire different shapes, forming the precursors of the plant tissue systems. Last, the apical meristems of the shoot and root systems are formed at the apical and basal ends of the embryo.

After an embryo has reached full size, developmental changes continue to occur at the cellular level. Embryonic cells, particularly those of the cotyledons, begin to synthesize and store the proteins, lipids, and starch that will provide the energy and basic building blocks for germination and seedling growth. Next, the embryo begins to desiccate, sometimes losing up to 80 percent of its previous water content, and enters a phase of dormancy. Development and metabolism are arrested in dormant embryos, and seeds containing dormant embryos can survive for many years (sometimes centuries) and withstand extreme temperatures and drought.

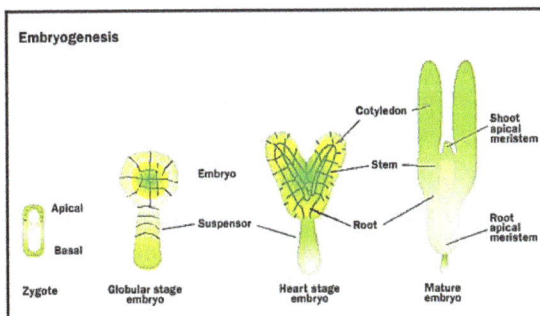

Embryo formation begins with cell division that establishes the apical-basal (top-bottom) axis. Further divisions elaborate on this basic plan, finally forming the cotyledons (seed leaves), as well as the apical meristems of root and shoot.

Plant hormones are important regulators of embryogenesis and seed dormancy. The hormones auxin, gibberellic acid, and cytokinin all stimulate growth and are present in the embryo during the stages of embryogenesis. As the embryo matures, these hormones are degraded and abscisic acid is synthesized by the embryo. Abscisic acid provides a developmental signal for the embryo to initiate the synthesis of storage compounds and to undergo desiccation. Abscisic acid is present in dormant seeds and is thought to play an important role in maintaining seed dormancy.

Germination and Seedling Development

Embryo development and metabolism resume upon seed germination. Given the right combination of water availability, temperatures, and light, the desiccated seed begins to take up water and the embryo begins to grow and metabolize again. Some species have specific requirements for germination; for instance, many temperate zone tree species require several weeks of temperatures of 4 degrees Celsius (39.2 degrees Fahrenheit) or less in order to germinate.

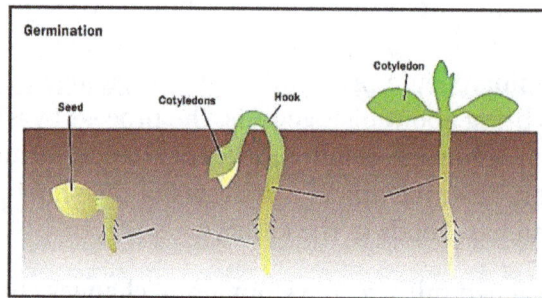

The root is the first portion of the plant to emerge during germination. Growth of the stem behind the cotyledons forms a "hook" that emerges from the soil, followed by emergence of the cotyledons, which begin to photosynthesize to feed further growth.

Other species require low levels of light in order to germinate. Once germination is initiated, the embryo follows a typical pattern of development. In many plants, the preformed embryonic root elongates first, forcing its way out of the seed coat and into the soil. Next, the embryonic stem, usually the part below the attachment of the cotyledons (the hypocotyl), elongates. Once the hypocotyl has carried the cotyledons into the light, they expand, providing a broad surface for photosynthesis.

Environmental factors and their translation into hormonal signals are important for seedling development. For instance, germination in the dark results in developmental events that help the seedling push its way through the soil into the light. The hypocotyl elongates quickly and maintains a "hook" near its tip that protects the cotyledons and shoot apical meristem region. Cotyledon expansion is suppressed so that they are not damaged as they are pulled through the soil. In contrast, if the same seeds germinate in the light, the hypocotyl hardly elongates at all and does not form a hook, while the cotyledons quickly expand. The hormone gibberellic acid plays an important role in seed germination and early seedling growth. Gibberellic acid induces the synthesis of enzymes required for the metabolism of stored foods, thus providing energy for seed-ling growth. Gibberellic acid also induces cell division and cell expansion in dark-grown hypocotyls, maintaining their rapid growth through the soil.

Apical Meristems and Development

The early stages of germination simply involve the enlargement of the root, hypocot-yl, and cotyledons that were preformed in the embryo. Postembryonic development, however, is focused on the apical meristems. The shoot apical meristem is the source of all the leaves, stems, and their component cells formed during the lifetime of the plant. The meristem itself is composed of a small population of perpetually embry-onic (meristematic) cells. These cells grow and divide; giving rise to new cells, but never matures themselves. Thus there is always a source of new cells at the tip of the shoot. The root tip has a similar population of meristematic cells that gives rise to all root tissues. Both of these meristems are characterized by an indeterminate growth pattern: one that is not finite, but, in theory at least, could continue throughout the lifetime of the plant.

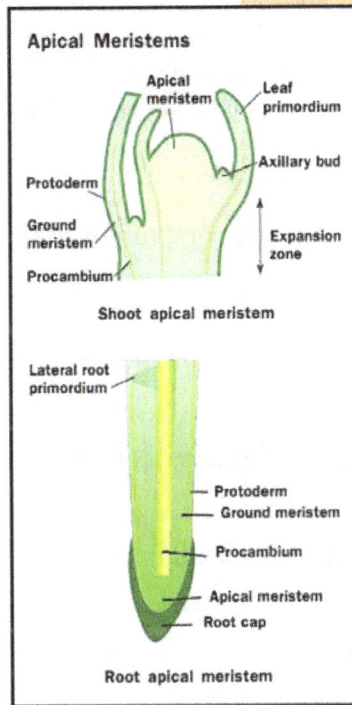

Structure of root and shoot apical meristems.

Apical meristems are involved in several distinct developmental processes. The mer-istems are the location of cell proliferation and thus the source of all new cells in the shoot and root systems. The regions below the meristems are the sites of active growth, as new shoot and root tissue rapidly expands. The shoot apical meristem plays a role in organogenesis, the formation of new leaves and axillary buds in a precise spatial pattern.

In contrast, the root apical meristem is not involved in organogenesis; lateral roots are initiated by pericycle cells, which are themselves derived from the meristem, usually several centimeters away from the meristem. The apical meristems also play a role in histogenesis by giving rise to cells that undergo distinct patterns of differentiation to form the specialized tissue types of the shoot and root. While the embryo initially gives rise to the precursors of dermal, ground, and vascular tissues (protoderm, ground meristem, and procambium, respectively), these tissue precursors continue to be formed by the apical meristems and represent the first stages of cell and tissue differentiation.

Cell Growth and Cell Division

Growth is defined as an irreversible increase in mass that is typically associated with an increase in volume. Plant cell growth is associated with meristems and must be carefully regulated in order for organogensis and histogenesis to occur in the appropriate patterns. The plant regulates growth by regulating the extensibility of its cell walls. A cell that has non-extensible cells walls can take up some water, but eventually the physical pressure of the water inside the cell pressing out on the cell wall (the turgor pressure) prevents the entry of additional water and any further change in volume. In contrast, a cell that has extensible cell walls can take up a substantial volume of water and thus increase in size. Turgor pressure that would otherwise prevent water entry momentarily decreases because the walls keep stretching.

Typically cell growth occurs in small increments:

- Wall extensibility increases, reducing turgor pressure;

- Reduced turgor pressure allows water to enter the cell, increasing cell volume;

- Wall extensibility decreases, allowing the cell to build up turgor and preventing further water entry; and

- The cell undergoes a cycle of synthesis of cytoplasmic and wall components, adding to the cell's mass.

This cycle of incremental growth is repeated many times until the cell reaches its final size.

The plant hormones auxin and gibberellin are produced in the vicinity of the apical meristems and usually act in concert to induce cell growth. Both hormones regulate wall extensibility, but carry out this function in different ways. Auxin induces the activity of cell membrane H^+ adenosine triphosphatase (ATPase) molecules. Proton (H^+) extrusion lowers the pH of the cell wall, thus activating the cell wall enzyme expansin. Expansin cleaves the hydrogen bonds between two cell wall components: The cellulose microfibrils and the hemicellulose molecules that link adjacent cellulose microfibrils. Breakage of these bonds allows these structural wall components to reposition themselves farther apart, increasing wall extensibility. Gibberellic acid, on the other hand, stimulates the activity of another cell wall enzyme called xyloglucan endotransglycosylase (XET). Xyloglucans are a type of hemicellulose that is cleaved by the XET enzyme. Breakage of the hemicellulose molecules also allows the cellulose microfibrils to move farther apart, increasing wall extensibility.

Cell division and cell growth are often tightly linked. When the rate of cell division is balanced by cell growth, as in the apical meristems, average cell size does not increase. As the meristem grows away from earlier formed cells, the ratio of growth to division increases, resulting in overall cell enlargement. As the tissues mature further, cell division ceases completely, giving rise to zones of pure cell enlargement where most of the visible growth of the plant occurs. This relationship between division and growth, coupled with observations of the predictable planes of cell division during histogenesis, indicates that cell division is carefully regulated during plant development.

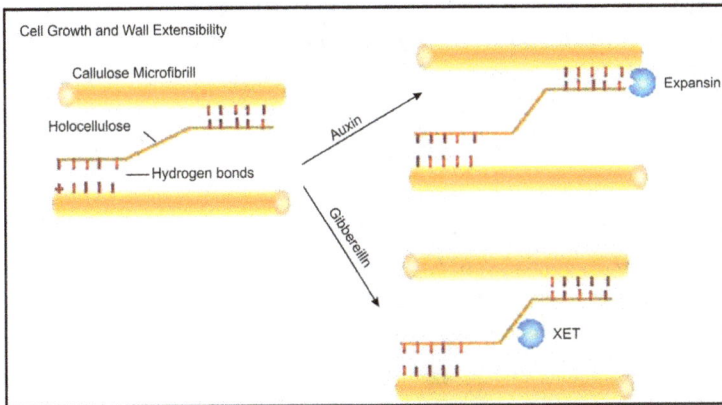

The hormones auxin and gibberellin each promote cell expansion by loosening the bonds between adjacent cell wall molecules. Each hormone acts on a different molecular target.

Molecules called cyclin-dependent kinases (CDKs) are key regulators of cell cycling (including cell division) in plants. CDKs are activated by association with a regulatory subunit called a cyclin and by phosphorylation and dephosphorylation events. The plant hormone cytokinin appears to regulate the cell cycle by interacting with the CDKs. Cytokinins enhance the synthesis of the cyclin subunits that are required for the cell to enter the deoxyribonucleic acid (DNA) synthesis phase of the cell cycle. Cytokinins also enhance the CDK dephosphorylation step that is required for the cell to progress into mitosis. Both of these processes are inhibited by the hormone abscisic acid; thus a "developmental tug-of-war" occurs between a division-enhancing hormone and a division-suppressing hormone. The delicate balance between them determines the rate of cell division and this type of interaction is probably typical of the hormonal regulation of many aspects of plant development.

Differentiation

Differentiation is the process whereby cells, tissues, and organs become different from each other and from their precursors. The concept can be applied to organogenesis since cotyledons, foliage leaves, sepals, and petals may all develop from similar appearing precursors, the leaf primordia. As these organs mature, they become different from each other in size, shape, and the development of distinctive cell types. For instance, the epidermis tissue of petals is sharply differentiated from that of cotyledons, foliage leaves and sepals that are photosynthetic organs. Correlated with a photosynthetic function, the epidermis of these organs is made up of flat, transparent cells that allow the penetration of light into internal tissues. Specialized guard cells that allow CO_2 to enter the leaf are also present. In contrast, the epidermal cells of petals contain brightly colored carotenoid or anthocyanin pigments. These cells also have a papillate shape that imparts a velvet like sheen to the petal surface. Since petals carry on minimal photosynthesis, they often lack guard cells.

The process of differentiation is best understood on a cellular level. For instance, guard cells are highly specialized epidermal cells. Early in the development of a leaf, protodermal precursor cells undergo a distinctive pattern of cell divisions. At first the cell divisions are asymmetric, producing one large and one small derivative. The large derivative stops dividing and differentiates as an unspecialized epidermal cell, while the small derivative undergoes another asymmetric division. At an unknown stop signal, the small derivative undergoes a symmetric division, giving rise to two equal sized cells that become the guard cells. Unlike their plain neighbors, these cells develop a distinctive kidney shape, unevenly thickened cell walls, large, conspicuous chloroplasts, and finally form a pore (the stomatal aperture) between them.

Uniqueness of Plant Development

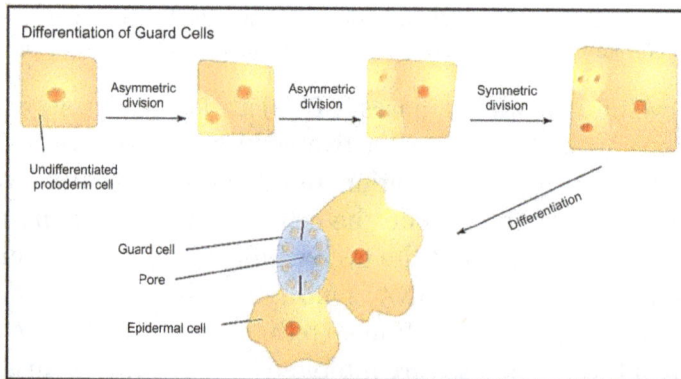

Guard cells regulate passage of gasses into and out of the leaf through pores in the surface. Guard cells form by a series of cell divisions from undifferentiated protoderm, including a final symmetric division that forms the two identical cells.

Although plants share many features of development with animals such as apical-basal polarity, regulation of the balance between cell growth and cell division, formation of distinctive patterns of organs, cells and tissues, and differentiation, some aspects of development are unique to plants. Among these are:

- The formation and maintenance of the perpetually embryonic regions, the apical meristems. The meristems have an indeterminate growth pattern that result in the occurrence of growth, organogenesis, and histogenesis throughout the life of the plant.

- Plant cells have rigid cell walls that prevent cell movement. Thus organogenesis and histogenesis must occur through differential growth and regulation of the planes of cell division. Cell-cell communication is important in plant development, but cell recognition is likely less important than it is in animals since plant cells keep the same neighbors throughout their life.

- Plant cells are totipotent; that is, able to differentiate as a different cell type if given the appropriate stimulus. Totipotency is likely a reflection of the plant's sedentary lifestyle. Plants can't escape predators and other kinds of damage, but they can readily repair wounds and reconnect vascular strands by differentiating the appropriate cell types.

Plant Morphology

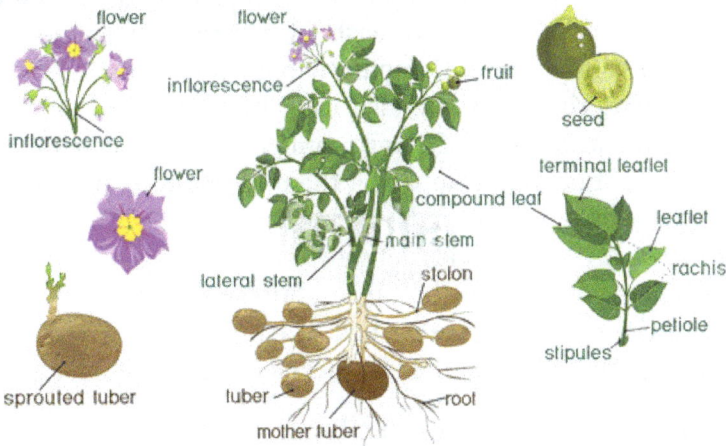

Potato plant

Plant morphology is the study of the structure and form development of plants in their individual and evolutionary-historical development. It is one of the most important branches of botany. As plant morphology developed, the following independent sciences were established: plant anatomy, which studies the tissue and cell structure of organs; plant embryology, which studies embryo development; and cytology, which studies cell structure and development. Thus, plant morphology, in the narrow sense, primarily studies the form development and structure of an organism; however, it also is concerned with regularities on the population-species level, inasmuch as it deals with the evolution of form.

Principal problems and methods: The principal concerns of plant morphology are identifying the morphological diversity of plants in nature, studying the regularities of the structure and interrelated distribution of organs and their systems, analyzing the changes in a plant's total structure and in its individual organs in the course of its independent development (onto morphogenesis), revealing the origins of plant organs during the evolution of the plant world (plant morphogenesis), and investigating the effects of various external and internal factors on form development. Thus, plant morphology does not limit itself to the description of certain types of structure but also strives to explain the dynamics of structures and their origins. The principles of biological organization, that is, the internal interrelationships of all processes and structures in an organism, are manifested in the organism's external form.

Theoretical plant morphology distinguishes between two interrelated and complementary approaches to the interpretation of morphological data. One approach reveals the reasons for the origin of any given form (from the point of view of the factors immediately acting on morphogenesis). The other explains the biological significance of certain forms for the life activity of the organisms (from the point of view of adaptability), which leads to the preservation of certain forms in the process of natural selection.

Morphological research is conducted by three principal methods: descriptive, comparative, and experimental. The first consists of descriptions of the forms of organs and their systems (organography). Comparative morphological methods are used to classify descriptive material. They are also used to study the changes an organism and its organs undergo as a result of aging (comparative-ontogenetic method), to explain the evolution of organs by comparing them with plants of different taxonomic groups (comparative-phylogenetic method), and to study the influence of the external environment (comparative-ecological method). Experimental morphological research involves the creation of controlled external conditions; the morphological reactions of plants to these artificial conditions are studied. The internal interrelationships between the organs of a living plant are examined by means of surgical intervention.

Plant morphology is closely related to other branches of botany, such as paleobotany, plant taxonomy and phylogeny (plant forms are the result of a long historical development and reflect the plants' similarities), plant physiology (the dependence of form on function), ecology, phytogeography and geobotany (dependence of form on external environment), genetics (inheritance and acquisition of new morphological characters), and horticulture.

Vegetative and Reproductive Characteristics

Plant morphology treats both the vegetative structures of plants, as well as the reproductive structures.

The vegetative (somatic) structures of vascular plants include two major organ systems:

(1) a shoot system, composed of stems and leaves, and (2) a root system. These two systems are common to nearly all vascular plants, and provide a unifying theme for the study of plant morphology.

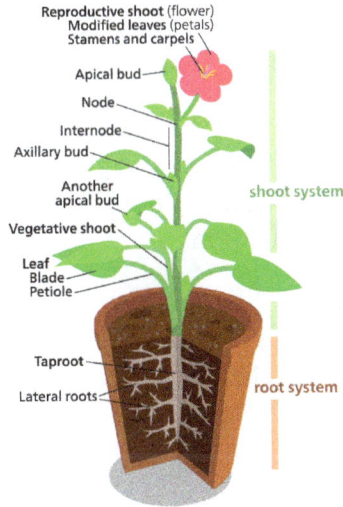

A diagram representing a "typical" eudicot.

By contrast, the reproductive structures are varied, and are usually specific to a particular group of plants. Structures such as flowers and fruits are only found in the angiosperms; sori are only found in ferns; and seed cones are only found in conifers and other gymnosperms. Reproductive characters are therefore regarded as more useful for the classification of plants than vegetative characters.

Use in Identification

Plant biologists use morphological characters of plants which can be compared, measured, counted and described to assess the differences or similarities in plant taxa and use these characters for plant identification, classification and descriptions.

When characters are used in descriptions or for identification they are called diagnostic or key characters, which can be either qualitative or quantitative.

- Quantitative characters are morphological features that can be counted or measured for example a plant species has flower petals 10–12 mm wide.

- Qualitative characters are morphological features such as leaf shape, flower color or pubescence.

Both kinds of characters can be very useful for the identification of plants.

Alternation of Generations

The detailed study of reproductive structures in plants led to the discovery of the alternation of generations, found in all plants and most algae, by the German botanist Wilhelm Hofmeister. This discovery is one of the most important made in all of plant morphology, since it provides a common basis for understanding the life cycle of all plants.

Pigmentation in Plants

The primary function of pigments in plants is photosynthesis, which uses the green pigment chlorophyll along with several red and yellow pigments that help to capture as much light energy as possible. Pigments are also an important factor in attracting insects to flowers to encourage pollination.

Plant pigments include a variety of different kinds of molecule, including porphyrins, carotenoids, anthocyanins and betalains. All biological pigments selectively absorb certain wavelengths of light while reflecting others. The light that is absorbed may be used by the plant to power chemical reactions, while the reflected wavelengths of light determine the color the pigment will appear to the eye.

Morphology in Development

Plant development is the process by which structures originate and mature as a plant grows. It is a subject studies in plant anatomy and plant physiology as well as plant morphology.

The process of development in plants is fundamentally different from that seen in vertebrate animals. When an animal embryo begins to develop, it will very early produce all of the body parts that it will ever have in its life. When the animal is born (or hatches from its egg), it has all its body parts and from that point will only grow larger and more mature. By contrast, plants constantly produce new tissues and structures throughout their life from meristems located at the tips of organs, or between mature tissues. Thus, a living plant always has embryonic tissues.

The properties of organization seen in a plant are emergent properties which are more than the sum of the individual parts. "The assembly of these tissues and functions into an integrated multicellular organism yields not only the characteristics of the separate parts and processes but also quite a new set of characteristics which would not have been predictable on the basis of examination of the separate parts." In other words, knowing everything about the molecules in a plant are not enough to predict characteristics of the cells; and knowing all the properties of the cells will not predict all the properties of a plant's structure.

Growth

A vascular plant begins from a single celled zygote, formed by fertilization of an egg cell by a sperm cell. From that point, it begins to divide to form a plant embryo through the process of embryogenesis. As this happens, the resulting cells will organize so that one end becomes the first root, while the other end forms the tip of the shoot. In seed plants, the embryo will develop one or more "seed leaves" (cotyledons). By the end of embryogenesis, the young plant will have all the parts necessary to begin in its life.

Once the embryo germinates from its seed or parent plant, it begins to produce additional organs (leaves, stems, and roots) through the process of organogenesis. New roots grow from root meristems located at the tip of the root, and new stems and leaves grow from shoot meristems located at the tip of the shoot. Branching occurs when small clumps of cells left behind by the meristem, and which have not yet undergone cellular differentiation to form a specialized tissue, begin to grow as the tip of a new root or shoot. Growth from any such meristem at the tip of a root or shoot is termed primary growth and results in the lengthening of that root or shoot. Secondary growth results in widening of a root or shoot from divisions of cells in a cambium.

In addition to growth by cell division, a plant may grow through cell elongation. This occurs when individual cells or groups of cells grow longer. Not all plant cells will grow to the same length. When cells on one side of a stem grow longer and faster than cells on the other side, the stem will bend to the side of the slower growing cells as a result. This directional growth can occur via a plant's response to a particular stimulus, such as light (phototropism), gravity (gravitropism), water, (hydrotropism), and physical contact (thigmotropism).

Plant growth and development are mediated by specific plant hormones and plant growth regulators (PGRs). Endogenous hormone levels are influenced by plant age, cold hardiness, dormancy, and other metabolic conditions; photoperiod, drought, temperature, and other external environmental conditions; and exogenous sources of PGRs, e.g., externally applied and of rhizospheric origin.

Morphological Variation

Plants exhibit natural variation in their form and structure. While all organisms vary from individual to individual, plants exhibit an additional type of variation. Within a single individual, parts are repeated which may differ in form and structure from other similar parts. This variation is most easily seen in the leaves of a plant, though other organs such as stems and flowers may show similar variation. There are three primary causes of this variation: positional effects, environmental effects, and juvenility.

Evolution of Plant Morphology

Transcription factors and transcriptional regulatory networks play key roles in plant morphogenesis and their evolution. During plant landing, many novel transcription factor families emerged and are preferentially wired into the networks of multicellular development, reproduction, and organ development, contributing to more complex morphogenesis of land plants.

Positional Effects

Although plants produce numerous copies of the same organ during their lives, not all copies of a particular organ will be identical. There is variation among the parts of a mature plant resulting from the relative position where the organ is produced. For example, along a new branch the leaves may vary in a consistent pattern along the branch. The form of leaves produced near the base of the branch will differ from leaves produced at the tip of the plant, and this difference is consistent from branch to branch on a given plant and in a given species. This difference persists after the leaves at both ends of the branch have matured, and is not the result of some leaves being younger than others.

Variation in leaves from the giant ragweed illustrating positional effects. The lobed leaves come from the base of the plant, while the unlobed leaves come from the top of the plant.

Environmental Effects

The way in which new structures mature as they are produced may be affected by the point in the plants life when they begin to develop, as well as by the environment to which the structures are exposed. This can be seen in aquatic plants and emergent plants.

Temperature

Temperature has a multiplicity of effects on plants depending on a variety of factors, including the size and condition of the plant and the temperature and duration of exposure. The smaller and more succulent the plant, the greater the susceptibility to damage or death from temperatures that are too high or too low. Temperature affects the rate of biochemical and physiological processes, rates generally (within limits) increasing with temperature. However, the Van't Hoff relationship for monomolecular reactions (which states that the velocity of a reaction is doubled or trebled by a temperature increase of $10°$ C) does not strictly hold for biological processes, especially at low and high temperatures.

When water freezes in plants, the consequences for the plant depend very much on whether the freezing occurs intracellularly (within cells) or outside cells in intercellular (extracellular) spaces. Intracellular freezing usually kills the cell regardless of the hardiness of the plant and its tissues. Intracellular freezing seldom occurs in nature, but moderate rates of decrease in temperature, e.g., 1° C to 6° C/hour, cause intercellular ice to form, and this "extra organ ice" may or may not be lethal, depending on the hardiness of the tissue.

At freezing temperatures, water in the intercellular spaces of plant tissues freezes first, though the water may remain unfrozen until temperatures fall below 7° C. After the initial formation of ice intercellularly, the cells shrink as water is lost to the segregated ice. The cells undergo freeze-drying, the dehydration being the basic cause of freezing injury.

The rate of cooling has been shown to influence the frost resistance of tissues, but the actual rate of freezing will depend not only on the cooling rate, but also on the degree of super cooling and the properties of the tissue. Sakai demonstrated ice segregation in shoot primordia of Alaskan white and black spruces when cooled slowly to 30° C to -40° C. These freeze-dehydrated buds survived immersion in liquid nitrogen when slowly rewarmed. Floral primordia responded similarly. Extra organ freezing in the primordia accounts for the ability of the hardiest of the boreal conifers to survive winters in regions when air temperatures often fall to -50° C or lower. The hardiness of the winter buds of such conifers is enhanced by the smallness of the buds, by the evolution of faster translocation of water, and an ability to tolerate intensive freeze dehydration. In boreal species of *Picea* and *Pinus*, the frost resistance of 1-year-old seedlings is on a par with mature plants, given similar states of dormancy.

Juvenility

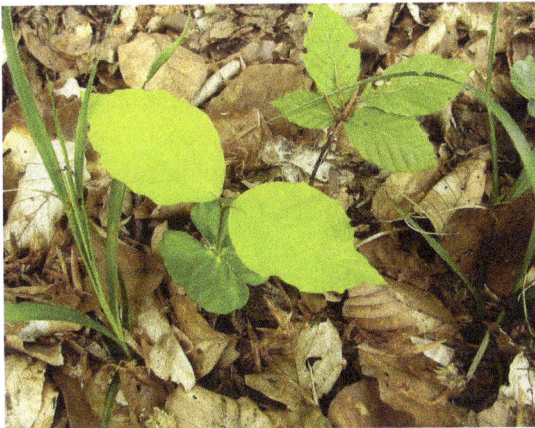

Juvenility in a seedling of European beech. There is a marked difference in shape between the first dark green "seed leaves" and the lighter second pair of leaves.

The organs and tissues produced by a young plant, such as a seedling, are often different from those that are produced by the same plant when it is older. This phenomenon is known as juvenility or heteroblasty. For example, young trees will produce longer, leaner branches that grow upwards more than the branches they will produce as a fully grown tree. In addition, leaves produced during early growth tend to be larger, thinner, and more irregular than leaves on the adult plant. Specimens of juvenile plants may look so completely different from adult plants of the same species that egg-laying insects do not recognize the plant as food for their young. Differences are seen in root ability and flowering and can be seen in the same mature tree. Juvenile cuttings taken from the base of a tree will form roots much more readily than cuttings originating from the mid to upper crown. Flowering close to the base of a tree is absent or less profuse than flowering in the higher branches especially when a young tree first reaches flowering age.

The transition from early to late growth forms is referred to as 'vegetative phase change', but there is some disagreement about terminology.

Modern Morphology

Rolf Sattler has revised fundamental concepts of comparative morphology such as the concept of homology. He emphasized that homology should also include partial homology and quantitative homology. This leads to a continuum morphology that demonstrates a continuum between the morphological categories of root, shoot, stem (caulome), leaf (phyllome), and hair (trichome). How intermediates between the categories are best described has been discussed by Bruce K. Kirchoff. A recent study conducted by Stalk Institute extracted coordinates corresponding to each plant's base and leaves in 3D space. When plants on the graph were placed according to their actual nutrient travel distances and total branch lengths, the plants fell almost perfectly on the Pareto curve. "This means the way plants grow their architectures also optimizes a very common network design tradeoff. Based on the environment and the species, the plant is selecting different ways to make tradeoffs for those particular environmental conditions."

Honoring Agnes Arber, author of the partial-shoot theory of the leaf, Rutishauser and Isler called the continuum approach Fuzzy Arberian Morphology (FAM). "Fuzzy" refers to fuzzy logic, "Arberian" to Agnes Arber. Rutishauser and Isler emphasized that this approach is not only supported by many morphological data but also by evidence from molecular genetics. More recent evidence from molecular genetics provides further support for continuum morphology. James concluded that "it is now widely accepted that radiality and dorsiventrality are but extremes of a continuous spectrum. In fact, it is simply the timing of the KNOX gene expression!." Eckardt and Baum concluded "it is now generally accepted that compound leaves express both leaf and shoot properties."

Process morphology (dynamic morphology) describes and analyzes the dynamic continuum of plant form. According to this approach, structures do not *have* processes, they *are* processes. Thus, the structure/process dichotomy is overcome by "an enlargement of our concept of 'structure' so as to include and recognize that in the living organism it is not merely a question of spatial structure with an 'activity' as something over or against it, but that the concrete organism is a spatio-*temporal* structure and that this spatio-temporal structure is the activity itself."

For Jeune, Barabé and Lacroix, classical morphology (that is, mainstream morphology, based on a qualitative homology concept implying mutually exclusive categories) and continuum morphology are sub-classes of the more encompassing process morphology (dynamic morphology).

Classical morphology, continuum morphology, and process morphology are highly relevant to plant evolution, especially the field of plant evolutionary biology (plant evo-devo) that tries to integrate plant morphology and plant molecular genetics. In a detailed case study on unusual morphologies, Rutishauser illustrated and discussed various topics of plant evo-devo such as the fuzziness (continuity) of morphological concepts, the lack of a one-to-one correspondence between structural categories and gene expression, the notion of morphospace, the adaptive value of bauplan features versus patio ludens, physiological adaptations, hopeful monsters and saltational evolution, the significance and limits of developmental robustness, etc.

Plant Embryogenesis

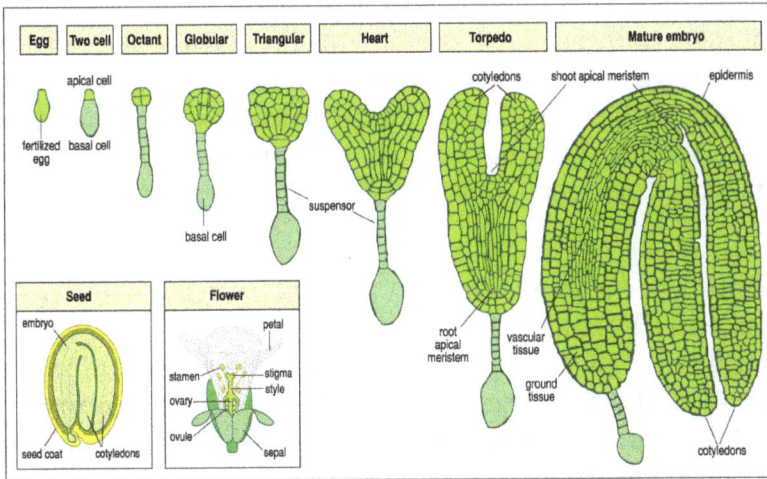

Embryogenesis is the process of initiation and development of an embryo from a zygote (zygotic embryogenesis) or a somatic cell (somatic embryogenesis). Embryo development occurs through an exceptionally organized sequence of cell division, enlargement and differentiation. Zygotic and somatic embryos share the same gross pattern of development. Both types of embryos develop as passing through typical developmental stages, such as globular, scutellar and coleoptilar stages for monocots, or globular, heart, torpedo and cotyledonary stages for dicots and conifers. Embryo development is bipolar, having a shoot and a radicular pole at opposite ends.

The physical, observable transition from a nonembryogenic cell to an embryogenic cell in somatic embryogenesis appears to occur when the progenitor cell undergoes an unequal division, resulting in a larger vacuolate cell and a small, densely cytoplasmic (embryogenic) cell. The embryogenic cell then either continues to divide irregularly to form a proembryonal complex or divides in a highly organized manner to form a somatic embryo.

The formation of the root apex involves derivatives of both the basal and apical cells of the two-cell embryo. The hypophyseal region, derived from the topmost cell of the suspensor, is incorporated into the embryo proper, giving rise to part of the root cap, its initial cells and the ground meristem initial cells. The remainder, including the ground meristem and procambium, is contributed by the apical cell.

The delineation of the embryonic shoot apex is much more cryptic than that of root apex. The shoot apical meristem, the epiphysis, is determined in an early globular stage embryo before the cell cleavage that delineates the protoderm. The storage protein mRNA are not detected in epiphyseal cells but in cotyledonary cells at one embryonic stage, indicating a functional difference, and the 'O' line is the lower boundary of the epiphysis. At the transition stage, the centrally localized epiphyseal cells divide more slowly than the cotyledon cell progenitors, resulting in the formation of a bilaterally symmetrical heart stage embryo.

Morphological changes during the transition from the globular stage to the heart stage are the first visible sign of the formation of the two embryonic organ systems: the cotyledons and the axis. The emergence of the cotyledons from a radically symmetrical globular embryo indicates that groups of cells in the apical region are induced to proliferate at specific sites. Polar auxin transport may be involved in directing these localized cell divisions.

Maturation is the terminal event of embryogenesis. In zygotic embryogenesis, maturation is characterized by attainment of mature embryo morphology, accumulation of storage carbohydrates, lipids and proteins, reduction in water content and a gradual decline or cessation of metabolism. Somatic embryos usually do not mature properly. Instead, due to environmental factors such as keeping a constant contact with inducing medium, somatic embryos often deviate from the normal developmental pattern by bypassing embryo maturation producing callus, undergoing direct secondary embryogenesis and/or germinating precociously. Somatic embryos growing from proembryonal complexes tend to develop asynchronously so that several stages are present in culture at any given time. Therefore, the most obvious developmental difference between zygotic and somatic embryos is perhaps that the latter lacks a quiescent resting phase.

Morphogenic Events

Embryogenesis occurs naturally as a result of single, or double fertilization, of the ovule, giving rise to two distinct structures: the plant embryo and the endosperm which go on to develop into a seed. The zygote goes through various cellular differentiations and divisions in order to produce a mature embryo. These morphogenic events form the basic cellular pattern for the development of the shoot-root body and the primary tissue layers; it also programs the regions of meristematic tissue formation. The following morphogenic events are only particular to eudicots, and not monocots.

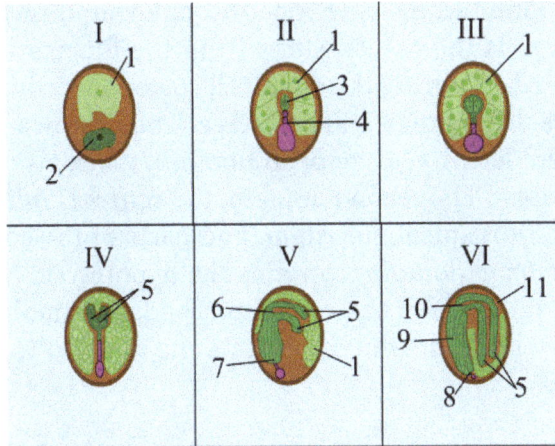

This depicts six different moments in embryogenesis.

I.	Two cell stage	1.	Endosperm
II.	Eight cell stage	2.	Single celled zygote
III.	Globular stage	3.	Embryo
IV.	Heart stage	4.	Suspensor
V.	Torpedo stage	5.	Cotyledons
VI.	Maturation	6.	Shoot apical meristem (SAM)
		7.	Root apical meristem (RAM)

Two Cell Stage

Following fertilization, the zygote and endosperm are present within the ovule, as seen in stage I of the illustration on this page. Then the zygote undergoes an asymmetric transverse cell division that gives rise to two cells - a small apical cell resting above a large basal cell. These two cells are very different, and give rise to different structures, establishing polarity in the embryo.

Apical Cell

The small apical cell is on the top and contains most of the cytoplasm, the aqueous substance found within cells, from the original zygote. It gives rise to the hypocotyl, shoot apical meristem, and cotyledons.

Basal Cell

The large basal cell is on the bottom and consists of a large vacuole and gives rise to the hypophysis and the suspensor.

Eight Cell Stage

After two rounds of longitudinal division, and one round of transverse division, an eight-celled embryo is the result. Stage II, in the illustration above, indicates what the embryo looks like during the eight cell stage. According to Laux, there are four distinct domains during the eight cell stage. The first two domains contribute to the embryo proper. The *apical embryo domain*, gives rise to the shoot apical meristem and cotyledons. The second domain, the *central embryo domain* \, gives rise to the hypocotyl, root apical meristem, and parts of the cotyledons. The third domain, the *basal embryo domain,* contains the hypophysis. The hypophysis will later give rise to the radicle and the root cap. The last domain, the *suspensor*, is the region at the very bottom, which connects the embryo to the endosperm for nutritional purposes.

Sixteen Cell Stage

Additional cell divisions occur, which leads to the sixteen cell stage. The four domains are still present, but they are more defined with the presence of more cells. The important aspect of this stage is the introduction of the protoderm, which is meristematic tissue that will give rise to the epidermis. The protoderm is the outermost layer of cells in the embryo proper.

Globular Stage

The name of this stage is indicative of the embryo's appearance at this point in embryogenesis; it is spherical or globular. Stage III, in the photograph above, depicts what the embryo looks like during the globular stage. 1 is indicating the location of the endosperm. The important component of the globular phase is the introduction of the rest of the primary meristematic tissue. The protoderm was already introduced during the sixteen cell stage. According to Evert and Eichhorn, the ground meristem and procambium are initiated during the globular stage. The ground meristem will go on to form the ground tissue, which includes the pith and cortex. The procambium will eventually form the vascular tissue, which includes the xylem and phloem.

Heart Stage

According to Evert and Eichhorn, the heart stage is a transition period where the cotyledons finally start to form and elongate. It is given this name in eudicots because most plants from this group have two cotyledons, giving the embryo a heart shaped appearance. Between the cotyledons is where the shoot apical meristem lies. Stage IV, in the illustration above, indicates what the embryo looks like at this point in development. 5 indicates the position of the cotyledons.

Torpedo Stage

This stage is defined by the continued growth of the cotyledons and axis elongation. In addition, programmed cell death must occur during this stage. This is carried out throughout the entire growth process, like any other development. However, in the torpedo stage of development, parts of the suspensor complex must be terminated. The suspensor complex is shortened because at this point in development most of the nutrition from the endosperm has been utilized, and there must be space for the mature embryo. After the suspensor complex is gone, the embryo is fully developed. Stage V, in the illustration above, indicates what the embryo looks like at this point in development.

Maturation

The second phase, or postembryonic development, involves the maturation of cells, which involves cell growth and the storage of macromolecules (such as oils, starches and proteins) required as a 'food and energy supply' during germination and seedling growth. The appearance of a mature embryo is seen in Stage VI above.

Dormancy

The end of embryogenesis is defined by an arrested development phase, or stop in growth. This phase usually coincides with a necessary component of growth called dormancy. Dormancy is a period in which a seed cannot germinate, even under optimal environmental conditions, until a specific requirement is met. Breaking dormancy, or finding the specific requirement of the seed, can be rather difficult. For example, a seed coat can be extremely thick. According to Evert and Eichhorn, very thick seed coats must undergo a process called scarification, in order to deteriorate the coating. In other cases, seeds must experience stratification. This process exposes the seed to certain environmental conditions, like cold or smoke, to break dormancy and initiate germination.

The Role of Auxin

Auxin is a hormone related to the elongation and regulation of plants. It also plays an important role in the establishment polarity with the plant embryo. Research, conducted by Cooke, Racusen, and Cohen, has shown that the hypocotyl from both gymnosperms and angiosperms show auxin transport to the root end of the embryo They hypothesized that the embryonic pattern is regulated by the auxin transport mechanism, and the polar positioning of cells within the ovule. The importance of auxin was shown, in their research, when carrot embryos, at different stages, were subjected to auxin transport inhibitors. The inhibitors that these carrots were subjected to made them unable to progress to later stages of embryogenesis. During the globular stage of embryogenesis, the embryos continued spherical expansion. In addition, oblong embryos continued axial growth, without the introduction of cotyledons. During the heart embryo stage of development there were additional growth axes on hypocotyls. Further auxin transport inhibition research, conducted on *Brassica juncea*, show that after germination, the cotyledons were fused and not two separate structures.

Alternative forms of Embryogenesis

Somatic Embryogenesis

Somatic embryos are formed from plant cells that are not normally involved in the development of embryos, i.e. ordinary plant tissue. No endosperm or seed coat is formed around a somatic embryo. Applications of this process include: clonal propagation of genetically uniform plant material; elimination of viruses; provision of source tissue for genetic transformation; generation of whole plants from single cells called protoplasts; development of synthetic seed technology. Cells derived from competent source tissue are cultured to form an undifferentiated mass of cells called a callus. Plant growth regulators in the tissue culture medium can be manipulated to induce callus formation and subsequently changed to induce embryos to form the callus. The ratio of different plant growth regulators required to induce callus or embryo formation varies with the type of plant. Asymmetrical cell division also seems to be important in the development of somatic embryos, and while failure to form the suspensor cell is lethal to zygotic embryos, it is not lethal for somatic embryos.

Androgenesis

The process of androgenesis allows a mature plant embryo to form from a reduced, or immature, pollen grain. Androgenesis usually occurs under stressful conditions. Embryos that result from this mechanism can germinate into fully functional plants. As mentioned, the embryo results from a single pollen grain. Pollen grains consists of three cells - one vegetative cell containg two generative cells. According to Maraschin et al., androgenesis must be triggered during the asymmetric division of microspores. However, once the vegetative cell starts to make starch and proteins, androgenesis can no longer occur. Maraschin et al., indicates that this mode of embryogenesis consists of three phases. The first phase is the *acquisition of embryonic potential*, which is the repression of gametophyte formation, so that the differentiation of cells can occur. Then during the *initiation of cell divisions*, multicellular structures begin to form, which are contained by the exine wall. The last step of androgenesis is *pattern formation*, where the embryo-like structures are released out of the exile wall, in order for pattern formation to continue.

After these three phases occur, the rest of the process falls in line with the standard embryogenesis events.

Plant Sporogenesis

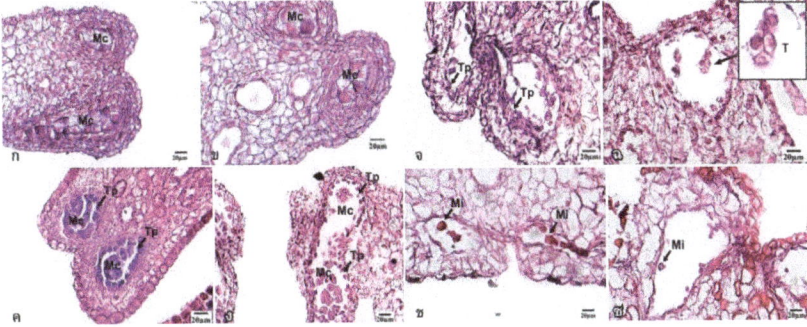

Sporogenesis is the formation of spores. In prokaryotes, that is plant organisms whose cells do not have typical nuclei, spores may arise from the entire cell, which has accumulated nutrient matter and thickened its capsule (for example, the exospores of many blue-green algae), or from a protoplast that has divided into a large number of spores (for example, the endospores of certain blue-green algae). Sporogenesis in prokaryotes may also result from the thickening and contraction of the protoplast inside the cell capsule and the formation of a new layered capsule on top of the protoplast (for example, in bacteria) or from the decomposition of special areas of mycelium into segments (for example, in actinomycetes).

Eukaryotes, that is, plants having typical nuclei, are characterized by three principal types of spores (oospores, mitospores, and meiospores) that occupy different places in the developmental cycle. Hence, there may be three variants of sporogenesis—oosporogenesis, mitosporogenesis, and meiosporogenesis, respectively. The term "sporogenesis" is usually used in reference to meiosporogenesis. Oosporogenesis is associated with fertilization and, consequently, with changes of nuclear phases in developmental cycles; it ends with the formation of oospores (in many green algae and oomycetes), auxospores (in diatoms), and zygospores (in zygomycetes), which consist of mononuclear or multinuclear zygotes. Mitosporogenesis leads to the development of mitospores, small or large numbers of which form as a result of mitotic divisions of haploid cells (for example, the zoospores of a number of algae and fungi) or, less frequently, diploid cells (for example, the carpospores of most Florideae). Mitospores sometimes form without cell division, for example, the monospores of Oedogonium, Bangiaceae, and Helminthocladiaceae.

A change in the nuclear phase does not occur in mitospores. It does occur, however, in unicellular mitosporangia (for example, in the zoosporangia of Ulothrix, the monosporangia of Oedogonium, and the cystocarps of Florideae). Unicellular algae seem to form sporangia themselves. Mitosporogenesis may accompany the decomposition of mycelium consisting of cells with dikaryons, for example, the mycelium of smut and rust fungi.

Meiosporogenesis is due to the replacement of diplophases by a haplophase in the developmental cycles of lower and higher plants. In lower plants meiospores arise during meiosis or shortly afterward from mitotically divided haploid cells formed during meiosis. In algae and fungi having a haploid cycle of development, sporogenesis involves the sprouting of the zygote (oospore), whose diploid nucleus forms four haploid nuclei by dividing meiotically. Four meiospores develop (for example, the zoospores of Chlamydomonas and the aplanospores of Ulothrix), or three out of the four haploid nuclei atrophy and only one mei-ospore is formed (for example, in Spirogyra).

Meiosis may be followed by one to three mitotic divisions, resulting in the formation of eight to 32 spores (for example, in Bangiaceae). In algae that have isomorphous or heteromorphous developmental cycles, meiosporogenesis occurs in unicellular meiosporangia and is characterized by the formation of only four meiospores (for example, the tetraspores of brown algae and most Florideae) or of 16 to 128 meiospores (for example, the zoospores of Laminariaceae) as a result of two to five mitotic divisions after meiosis. In the sporangia of ascomycetous fungi (sacs, or asci) the four haploid nuclei resulting from meiosis divide mitotically, and eight endogenous meiospores (ascospores) are formed. In basidiomycetous fungi, four haploid nuclei arise in each of the sporiferous organs, or basidia. The four nuclei move to special outgrowths on the surfaces of the basidia, and the outgrowths, known as basidiospores, subsequently separate from the basidia.

Higher plants form only meiospores; meiosporogenesis occurs in multicellular sporangia. Sporocytes (meiotically dividing cells) usually develop as a result of mitotic divisions of the diploid cells of the archespore; they each form four spores (tetrads of spores). Isosporous pteridophytes produce morphologically and physiologically identical spores, from which bisexual prothallia develop. In heterosporous pteridophytes and seed plants two types of spores develop as a result of microsporogenesis, megasporogenesis, and meiosporogenesis.

Microsporogenesis occurs in the microsporangia and is completed by the formation of a large number of microspores, which subsequently develop into male prothallia. Megasporogenesis occurs in the megasporangia, where a small number of mega-spores—often only four or one—mature and develop into female prothallia. The developing sporocytes and spores (in most higher plants) feed on substances obtained from cells of the tapetum, that is, the layer that lines the interior of the cavity of the sporangium. In many plants the tapetal cells dissolve and form a peri-plasmodium (a protoplasmic mass with degenerating nuclei), which contains sporocytes and, later, spores. In a number of plants some of the sporocytes participate in formation of the periplasmodium.

In the megasporangia, or ovules, of some angiosperms, meiosis results in formation of cells having two or four haploid nuclei, which correspond to two or four megaspores. Female gametophytes, in the form of bisporous and tetrasporous embryo sacs, develop from these cells.

Plant Megasporogenesis and Megagametogenesis

The ovule develops as multicellular placental outgrowth including the epidermal and a number of hypodermal cells. With further development, this gives rise to nucellus and one or two integuments from its basal region. In ovules, with two integuments, usually the inner one is formed first than the outer one. The inner one is more delicate and inconspicuously developed than the outer one.

One hypodermal cell of the nucellus becomes differentiated from the other by its bigger size, dense cytoplasm and conspicuous nucleus, called archesporial cell. The archesporial cell divides transversely and forms an inner primary sporogenous cell and an outer primary parietal cell.

The primary sporogenous cell functions as megaspore mother cell and the primary parietal cell undergoes repeated vertical divisions and forms layers of parietal cells. Sometimes, the archesporial cell does not divide and directly functions as megaspore mother cell.

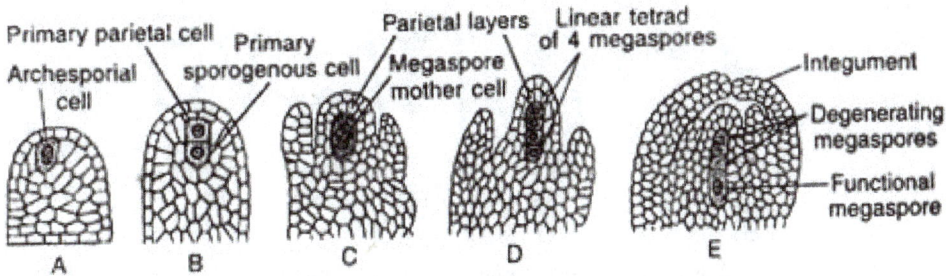

A-E. Stages of development of megaspore mother cell and megasporogenesis

Megasporogenesis (Development of Megaspores)

The megaspore mother cell is diploid (2n), which undergoes meiosis and forms four haploid (n) megaspores. The first division of megaspore mother cell is transverse, forming two cells. Both the cells again divide transversely and thus four haploid megaspores are formed.

The megaspores are then arranged in an axial row, called linear tetrad. Out of four megaspores, only one which remains towards the chalazal end behaves as functional megaspore and the other three which remain towards the micropylar end, gradually degenerate. The functional megaspore forms the female gametophyte i.e., the embryo sac.

Megagametogenesis (Formation of Female Gametophyte i.e., Embryo Sac)

Megaspore (n) is the first cell of the female gametophyte. The functional megaspore becomes enlarged at the expense of tape tum and the nucellus and thus forms the female gametophyte i.e., the embryo sac. Initially, the embryo sac is uninucleate and with further growth its nucleus divides by three successive divisions and forms eight nuclei in figure.

Out of eight nuclei, initially four remain towards the micropyle end and the other four towards the chalazal end. One nucleus from each pole then moves towards the center and forms a pair of polar nuclei. These nuclei fuse together and form 2n nucleus, the definitive nucleus. It is also known as polar fusion nucleus or secondary nucleus.

The three nuclei of the micropylar end form the egg apparatus and the rest three at the chalazal end are called antipodal cells. In the egg apparatus, each nucleus is surrounded by viscous mass of cytoplasm without any wall, of which the middle one is the largest and called egg, ovum or oosphere and the rest two (one on each side of the egg) are the synergids or helping cells. The antipodal cells have viscous mass of cytoplasm, covered by cellulosic wall.

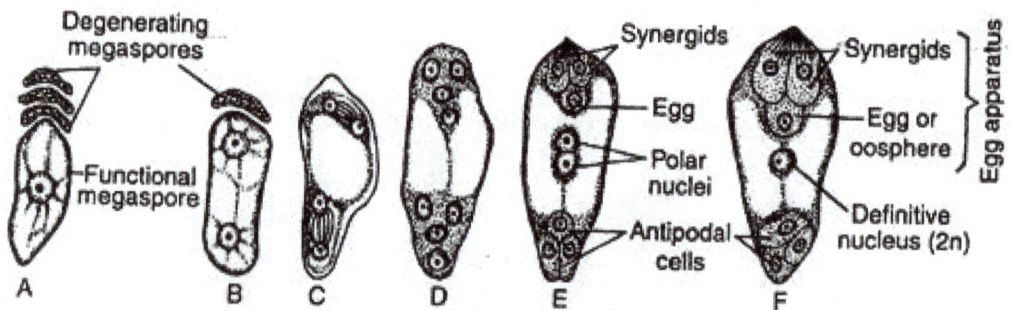

A-F. stages of development of female gametophyte

This type of embryo sac development is very common in angiosperms and is known as ordinary type or normal type or Polygonum type. This type is also known as monosporic type, because, out of four megaspores, only one remains functional and forms the embryo sac.

Other Types of Embryo Sac Development:

Monosporic Type

Oenothera type: In this type (like Polygonum type), usual linear tetrad of megaspores are formed, but instead of the innermost one, the outermost megaspore (which is present towards micropyle) remains functional and forms the embryo sac. The functional megaspore undergoes two successive divisions and forms 4 nuclei.

All the nuclei remain towards the micropyle. Out of four nuclei, three nuclei form the egg apparatus (egg and two synergids) and the remaining one forms a single polar nucleus. Second polar nucleus and antipodal cells are absent, e.g., Oenothera and other members of Onagraceae.

Bisporic Type

Allium type: The megaspore mother cell divides to form two cells, the upper one quickly degenerates. The lower one then divides and forms two nuclei, distributed in the two poles. Later on, both the nuclei undergo two successive divisions and form usual octant type of embryo sac, i.e., polygonum type. Here two megaspore nuclei take part in the development of embryo sac i.e., bisporic type, e.g., Allium, Scilla, Trillium etc., of Liliaceae.

Tetrasporic Type

Peperomia type: The megaspore mother nucleus undergoes meiotic division and forms four nuclei, which remain crosswise in the embryo sac without any wall. All the nuclei undergo two successive divisions and form 16 nuclei which remain dispersed inside the sac. Later on, out of 16 nuclei, egg and one synergid remain at the micropylar end, six antipodal cells towards the chalazal end, and the rest eight at the center forming polar nuclei, e.g., Peperomia of Piperaceae etc.

Penaea type: Like Peperomia type, 16 nuclei are formed, those remain crosswise in the embryo sac. Later on, the nuclei are distributed in a different manner. The egg and two synergids remain at the micropylar end, three nuclei at the chalazal end, and four at the centre and three each on the two side walls, e.g., Penaea of Penaeaceae.

Drusa type: Like Peperomia type, initially four megaspores are formed, these are distributed in different ways. One megaspore remains towards the micropyle, and the rest three at the chalazal end.

All the nuclei undergo two divisions and form 16 nuclei, out of which four nuclei remain towards the micropyle and the rest twelve at the chalazal end. In the mature embryo sac, egg and two synergids remain towards the micropyle, two (one from each pole) at the centre and the rest eleven at the chalazal end, e.g., Drusa oppositifolia of Apiaceae.

Fritillaria type: Like Drusa type, out of four nuclei formed, one nucleus remains towards the micropyle, and the rest three at the chalazal end. The chalazal nuclei fused together and form 3n nucleus. Both the cells thus undergo one mitotic division and again form a tetrasporic stage. Out of four nuclei, two remain at each pole.

All the nuclei then undergo mitotic division and form eight nuclei. Out of four haploid nuclei at the micropyle, one egg and two synergids are formed, those remain at the micropylal end; three triploid nuclei at the chalazal end and one from each pole remain at the centre (one haploid and the other one triploid), e.g., Fritillaria, Tulipa and some other members of Liliaceae.

TYPE	MEGASPOROGENESIS			MEGAGAMETOGENESIS			
	Megaspore mother cell	Division I	Division II	Division III	Division IV	Division V	Mature embryo sac
Monosporic 8-nucleate Polygonum type							
Monosporic 4-nucleate Oenothera type							
Bisporic 8-nucleate Allium type							
Tetrasporic 16-nucleate Peperomia type							
Tetrasporic 16-nucleate Penaea type							
Tetrasporic 16-nucleate Drusa type							
Tetrasporic 8-nucleate Fritillaria type							
Tetrasporic 4-nucleate Plumbagella type							
Tetrasporic 8-nucleate Plumbago type							
Tetrasporic 8-nucleate Adoxa type							

Figure: Development of different types of embryo sac in angiosperms

Plumbagella type: It is like Fritillaria type, which forms 1st and 2nd tetrasporic stage with two haploid nuclei at the micropyle and two triploid nuclei at the chalazal end of the embryo sac. Later on, the nuclei are distributed in such a way that the egg is at the micropyle, one triploid nucleus at the chalazal end and one triploid plus one haploid nuclei at the centre, e.g., Plumbagella of Plumbagellaceae.

Plumbago type: It is like Penaea type where firstly four nuclei are formed followed by eight nucleated embryo sac. The two nuclei at each side (four sides) remain crosswise.

Later on, four nuclei, one from each side, become aggregated in the center. The nucleus at the micropylar end behaves as egg, e.g., Plumbago of Plumbaginaceae.

Adoxa type: In this type, the megaspore mother nucleus divides meiotically into four nuclei arranged two at each end. Both the nuclei — further undergo mitotic division and thus eight nuclei are formed. Like the normal type i.e., Polygonum type, one egg and two synergids remain at the micropylar region, three antipodal cells at the chalazal end and two nuclei remain in the center, e.g., Adoxa, Sambucus of Caprifoliaceae.

Epigenetics of Plant Growth and Development

Epigenetic is one of the most important topics in the field of plant genetics. It is a promising aspect to impart plant stress in different plant species. Many scientific studies have been supporting the development of plant genetics. Epigenetic is an important aspect to solve the problems in transgenic plants, with suitable expression from new transgenic segments. In this paper some of the important aspects of epigenetic are enlighten to support the future research studies. Plants gained the ability to change their response to environmental stimuli. Epigenetic changes in gene expression have fascinated scientists over several decades. These processes received particular attention in plants, where they can result in beautiful variations in conspicuous phenotypes such as pigmentation. Epigenetic control is also a key issue in the development of transgenic plants with appropriate expression from newly introduced transgene segments. The term 'epigenetic' refers to heritable gene expression patterns determined by how the DNA of a gene is packaged rather than its primary DNA sequence. Genes are tightly packed within DNA and they are not available to the transcription machinery and are expressed very poor. Normally the patterns of DNA packaging are carefully controlled to give predictable patterns of gene expression. However, the process can occasionally go awry to cause altered gene expression. This primer will focus on well-characterized examples of epigenetic changes in plants that shed light on the mechanisms underlying this fundamental gene control process. Defense, plants evolved sophisticated mechanisms to respond and acclimatize to these stresses by prompt and harmonized changes at transcriptional and post-transcriptional levels of whole gene complexes. According to Richards, the range of epigenetic variation in relation to genotypic context could be categorized as three: obligatory, facilitated and pure. Whereas the obligate epigenetic variation is completely dependent on genetic variation, facilitated epigenetic variation is semi-independent of genetic variation, and pure epigenetic variation is completely independent of genetic variation.

Role of Epigenetics and Plant Genetics

The effect of epigenetic over the plants and crops such as maize has many strains with striking patterns of kernel or plant pigmentation was selected for cultivation. These strains were provided a rich source of epigenetic variation in pigment gene expression. One such case that has been examined at the molecular level is the expression of a transcription factor gene that controls pigment synthesis. According to Judith Bender the variations in epigenetic results to pigment gene expression. Cotton is not only the most important source of renewal textile fibers, but also an important model for studying cell fate determination and polyploidy effects on gene expression and evolution of domestication traits. The combination of A and D-progenitor genomes into allotetraploid cotton induces inter genomic interactions and epigenetic effects, leading to the unequal expression of homologous genes. Small RNAs regulate the expression of transcription and signaling factors related to cellular growth, development and adaptation.

Epigenetic changes in DNA and chromatin affect the genes and transposons activity. Epigenetic controls the time of flowering, parent of origin imprinting, paramutation and transposon silencing. Genome studies of epigenetic marks revealed the critical role of small interfering RNAs in maintaining epigenetic states.

Tomato is an important vegetable all over the world and it is important to carry research regarding it, to develop its growth. A novel insight into gene regulatory mechanism was presented by an epigenetic modification study of a tomatoprotein-coding non-transposon ASR1 (Abscisic acid stress, ripening 1) epiallele, demonstrating its role in DNA methylation during water deficit stress. According to Gonzalez et al. DNA methylation was occurred with possible changes in epigenetics of tomato.

From the past few years there was an increase in the modification of agriculture crops with transgenes, which express desirable traits. However, a frequent stumbling block is the unwanted silencing of the transgene. Many studies of silenced transgenes and silenced endogenous sequences showed that repeated sequence arrays, in particular inverted repeats are most likely silencing. Thus, selection of transgenic plants with single copy transgene insertions is the first line of defense against silencing. An interesting approach to identify epigenetically controlled probable gene sets in response to drought stress was performed by Shaik and Ramakrishna. Epigenetics and plant genetics are the most important areas to develop agriculture and the plant development.

Seed Dormancy and Germination

Germination is the early growth of a plant from a seed. Meanwhile, dormancy precedes germination and serves to preserve a seed until conditions are receptive towards growth. The transition from dormancy to germination seems to depend on the removal of factors inhibiting growth. There are many models for germination which may differ between species. The activity of genes such as Delay of Germination and presence of hormones such as gibberelins have been implicated in dormancy while the exact mechanisms surrounding their action is unknown.

Deacetylation

There are at least eighteen histone deacetylases in *Arabidopsis*. Genome-wide association mapping has shown that deacetylation of histones by Histone Deacetylase 2B negatively affects dormancy. The remodeling of chromatin by histone deacetylase leads to silencing of genes that control plant hormones such as ethylene, abscisic acid, and gibberelin which maintain dormancy. Additionally, Histone Deacetylase A6 and A19 activity contributes to silencing of Cytochrome P450 707A and activation of NINE-CIS-EPOXYCAROTENOID DIOXYGENASE 4, 9. Both of these actions lead to increased abscisic acid.

Methylaltion

Methylation by the methyltransferase KRYPTONITE causes histone H3 lysine 9 dimethylation which recruits the DNA methyltransferase CHROMOMETHTLASE3 in tandem with HETEROCHROMATIN PROTEIN1. This association methylates cytosine for a stable silencing of Delay of Germination 1 and ABA Insensitive Genes which both contribute to dormancy.

Flowering and Related Mechanisms

Flowering is a pivotal step in plant development. Numerous epigenetic factors contribute to the regulation of flowering genes, known as flowering loci (FL). In *Arabidopsis*, flowering locus t is responsible for the production of florigen, which induces changes in the shoot apical meristem, a special set of growth tissues, to establish flowering. Homologs of the flowering genes exist in flowering plants, but the exact nature of how the genes respond to each mechanism might differ between species.

Vernalization

Vernalization depends on the presence of a long non-coding RNA that is termed COLDAIR. The exposure of plants to a significant period of cold results in COLDAIR accumulation. COLDAIR targets polycomb repressive complex 2 which acts to silence flowering locus C through methylation. As flowering locus C is repressed it no longer acts to inhibit the transcription of flowering locus t and SOC1. Flowering locus t and SOC1 activity leads to the development of flowers.

Photoperiodism

Another set of flowering controls stems from photoperiodism which initiates flowering based on the length of nighttime. Long day plants flower with a short night, while short day plants require uninterrupted darkness. Some plants are restricted to either condition, while others can operate under a combination of the two, and some plants do not operate under photoperiodism. In *Arabidopsis*, the gene CONSTANS responds to long day conditions and enables flowering when it stops repressing flowering locus t. In rice, photoperiodic response is slightly more complex and is controlled by the florigen genes Rice Flowering locus T 1 (RFT1) and Heading date 3 a (Hd3a). Hd3a, is a homolog of flowering locus t and, when no longer repressed, activates flowering by directing modification of DNA at the shoot apical meristem with florigen. Heading date 1 (Hd1) is a gene that promotes flowering under short day conditions but represses flowering under long day as it either activates or suppresses Hd3a. Meanwhile, RFT1 can cause flowering under non-inductive long day. Polycomb Repressive Complex 2 can lead to silencing of the genes through histone H3 lysine 27 trimethylation. A variety of chromatin modifications operating in both long and short days or only under one condition can also affect the two florigen genes in rice.

Flowering Wageningen (FWA)

The gene FWA has been identified as responsible for late flowering in epi-mutants. Epi-mutants are individuals with particular epigenetic changes that lead to a distinct phenotype. As such, both wild type and epi-mutant variants contain identical sequences for FWA. Loss of methylation in direct repeats in the 5' region of the gene results in expression of FWA and subsequent prevention of proper flowering. The gene is normally silenced by methylation of DNA in tissues not related to flowering.

Meristematic Tissue

Meristematic tissues contain cells that continue to grow and differentiate throughout the plant's lifetime. Shoot apical meristem gives rise to flowers and leaves while root apical meristem grows into roots. These components are crucial to general plant growth and are the harbingers of development. Meristematic tissue apparently contains characteristic epigenetic modifications. For example, the boundary between the proximal meristem and elongation zone showed elevated H4K5ac along with a high level of 5mC in barley. Root meristematic tissues have been found to contain patterns for histone H4 lysine 5 acetylation, histone H3 lysine 4 and 9 di methylation and DNA methylation as 5-methyl cytosine. So far, only a casual correlation between epigenetic marks and tissue types has been established and further study is required to understand the exact involvement of the marks.

Heterosis

Heterosis is defined as any advantages seen in hybrids. The effects of heterosis seem to follow a rather simple epigenetic premise in plants. In hybrids, lack of proper regulatory action, such as silencing by methylation, leads to uninhibited genes. If the gene is involved in growth, such as photosynthesis, the plant will experience increased vitality.

The results heterosis can be seen in traits such as increased fruit yield, earlier ripening, and heat tolerance. Heterosis has been shown to provide increased general growth and fruit yield in tomato plants (Xiaodong).

Plant Photomorphogenesis

Light is an important environmental factor which controls growth and development in plants. Besides photosynthesis in which light is harvested by green plants and is convert-ed- into chemical energy, there are numerous other plant responses to light such as pho-totropism, germination of some light sensitive seeds e.g. lettuce, de-etiolation of mono-cot and dicot seedlings etc., which are quite independent of photosynthesis and in which light just acts as environmental signal to bring about the particular photo-response.

Most of these photo-responses control genetically defined structural development or morphogenesis (i.e., origin of form) of plants. The role of light in regulating morpho-genesis is known as photo-morphogenesis. In plants, red and blue light are especially effective in inducing a photo-morphogenetic response.

The effect of light in controlling morphogenesis can best be demonstrated by compar-ing a monocot (maize) or dicot (bean) seedling grown in light with one grown in dark-ness both of which have been reared from genetically identical seeds. Abundant reserve food in seeds eliminate the need for photosynthesis for many days.

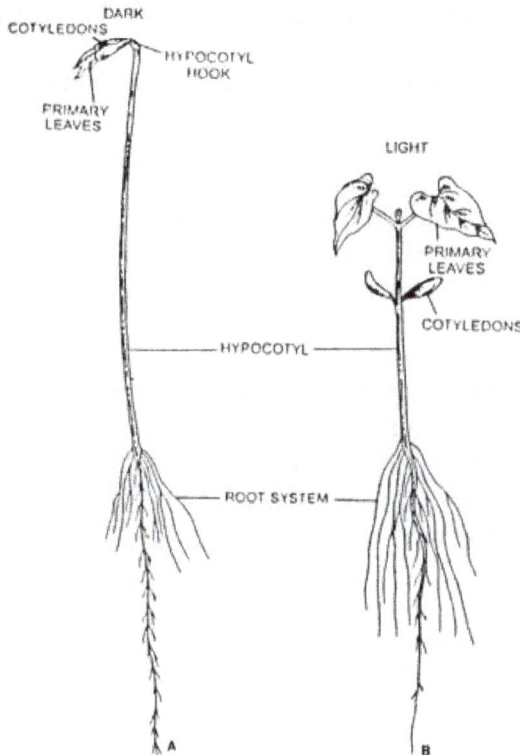

Figure: Effect of light on seeding development in bush bean.

It can easily be noticed that dark grown seedling has become etiolated (i.e., pale and weak) while the one grown in light has stockier and green appearance with short stem and large leaf area. Since both etiolated and light grown seedlings were reared from genetically identical seeds, light must have altered the gene expression during germination so that the appearance or form of etiolated and light grown seedlings looks different.

De-etiolation of light grown seedling can be done in very short period (hours) by placing it even in dim light. During de-etiolation, marked reduction in the rate of stem elongation, straightening of apical hook and development of green pigments can easily be noticed. The etiolated form of the seedling is thus gradually transformed to stockier green appearance and is the result of photo-morphogenesis. The development of seedling in darkness is called as skoto-morphogenesis.

According to Hans Mohr, there are two important stages of photo-morphogenesis:

- Pattern specification, in which cells and tissues develop specific ability or competence to respond to light during certain developmental stage, and

- Pattern realization, during which time the photo-response occurs.

There are two main categories of plant responses to light signals:

- Phytochrome mediated photoresponses and

- Blue-light responses or cryptochrome mediated photo-responses.

(A) Phytochrome Mediated Photoresponses in Plants

Large number of photo-morphogenic responses in plants are known to be mediated by the proteinaceons pigment (chromoprotein) called phytochrome. This pigment acts as photoreceptor and absorbs most strongly red and far-red light. It also absorbs blue light.

The pigment phytochrome exists in two forms, (i) a red light absorbing form designated as P_R form and (ii) another far-red light absorbing form designated as P_{FR} form. These two forms are photochemically interconvertible. When P_R form absorbs red light (650 – 680 nm), it is converted to P_{FR} form. The P_{FR} form absorbs far-red light (710 – 740 nm) and is converted to P_R form. The P_{FR} form of this pigment is believed to be physiologically active form.

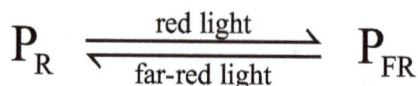

$$P_R \xrightleftharpoons[\text{far-red light}]{\text{red light}} P_{FR}$$

Absorption spectra of P_R and P_{FR} forms of phytochrome purified from etiolated Avena seedlings are given in figure. P_R form shows a peak at 666 nm while P_{FR} form at 730 nm. It is noteworthy that both forms of phytochrome also absorb in the blue region of spectrum of light.

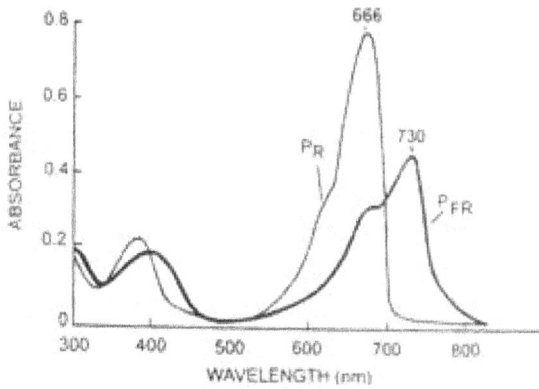

Absorption spectra of purified phytochrome in P_R and P_{FR} form etiolated Avena seedlings.

(The absorption maxima obtained in vitro mostly correspond with those in vivo provided that the native phytochrome is carefully purified and non-degraded).

However, it is quite obvious from this figure that the absorption maxima of P_R and P_{FR} forms overlap considerably in the red region of the spectrum of visible light and therefore, the phytochrome system cannot be quantitatively converted from P_R to P_{FR}. After irradiation with red light (or white light), there is an equilibrium between P_R and P_{FR} forms that depends upon the spectral composition of the light source. This equilibrium is called as photo-stationary equilibrium (φ) and is defined as ratio of the P_{FR} conc. and the total phytochrome conc. (ρ_{total}) at a given wavelength. These values can be measured by difference spectroscopy.

$$\phi = P_{FR}/ P_R + P_{FR} = P_{FR}/P_{total}$$

Most of the phytochrome mediated photoresponses in plants are reversible. These are induced by red light and reversed by far red light. A list of some of the photo reversible responses mediated by phytochrome in plants is given in table.

Plants group	plant	Developmental stage	Morphogenic response to red light
Angiosperms	Lettuce (lactuca sativa)	Seed	Germination of seeds
	Oat (Avena sativa)	Etiolated seedling	De-etionlation
	Cocklebur (Xanthium)	Adult	Inhibitions of flowering
	Mustard (Sinapsis)	Seedling	Stimulates formations of leaf primordia, development of primary leaves and production of anthocyanin.
Gymnosperms	Pine (Pinus)	Seedling	Stimulates rate of chlorophyll accmulation
Pteridophytes	(Sensitive fern) Onoclea	Young gametophy	Stimulates growth

| bryophytes | Moss (Polytrichum) | sporeling | Stimulates replicationof plastids |
| Green algae | Mougeotia | Adult gametophyte | Rotation of chloroplast |

Table: Some typical photoreversible responses mediated by phytochrome in plants.

One of the classical examples of photo-morphogenesis in plants induced by short red-far- red pulses is the germination of light sensitive seeds of lettuce (Lactuca sativa). In early 1930s, Flint and McAlister demonstrated that germination of lettuce seeds is not only stimulated by white light but also by red light (shorter than 700 nm) and inhibited by far-red light (greater than 700 nm).

In 1950s, Borthwick and Hendricks and their associates obtained spectacular results by exposing lettuce seeds to alternating red and far-red treatments. They observed much higher percentage of germination when the seeds received red light as the final treatment. Seed germination was markedly inhibited when seeds received final treatment with far-red light in table.

Red/Far Red (R/FR) Light Treatment	Percentage Germination
R (5 min.)	70%
R+FR (5 min. each)	6%
R+FR+R	74%
R+FR+R+FR	6%
R+FR+R+FR+R	76%
R+FR+R+FR+R+FR	7%
R+FR+R+FR+R+FR+R	81%
R+FR+ R+FR+ R+FR+ R+FR	7%

Table: Effect of alternating red/far red (R/FR) light treatment on germination of lettuce seeds

Borthwick and his associates also predicted existence of the photoreceptor phyto-chrome in two different forms which was proved to be absolutely correct later on when this pigment was isolated in plant extracts for the first time by Butler et al in 1959 and its photo-reversibility was confirmed in vitro.

Based on the amount of light required or the fluence (no. of photons absorbed per unit surface area), the phytochrome mediated photo-responses can be grouped into three main categories:

- Very Low Fluence Responses (VLFRs)

  These responses are initiated by very low fluences (0.1 to 1 n mol m^{-2}) saturating at 50 n mol m^{-2} and are non-photo reversible. For example, brief flash of red light with fluence as low as 0.1 n mol m^{-2} can stimulate the growth of coleoptile and inhibit growth of mesocotyl in oat seedlings that have been grown in dark. Similarly, red light with fluence of only 1-100 n mol m^{-2} is enough to stimulate seed germination in Arabidopsis. (In monocots, the elongated area of axis between coleoptile and root is called as mesocotyl)

- Low Fluence Responses (LFRs)

These responses require fluence of at least 1.0 nmol m^{-2} saturating at 1000 n mol m^{-2} and are photo-reversible. Most of the red/far-red photo-responses including the lettuce seed germination belong to this category.

- High Irradiance Responses (HIRs)

These responses require continuous or prolonged exposure to light of relatively high irradiance saturating at much higher fluences (at least 100 times more) than LFRs and are non- photo-reversible.

Examples are:
- Anthocyanin synthesis in dicot seedlings and in apple skin,
- Ethylene production in sorghum,
- Induction of flowering in Hyoscyamus (a long day plant),
- Opening of plumular hook in lettuce,
- Enlargement of cotyledons in mustard,
- Inhibition of hypocotyl elongation in many dicot seedlings etc.

(B) Blue Light Responses or Cryptochrome Mediated Photoresponses

Apart from phytochrome mediated photo-responses, large number of photo-responses in plants are known which are controlled by blue light and are believed to be mediated through a group of yet unidentified pigments called crypto chrome (crypto from cryptogams), the latter acting as photoreceptor in such responses. Blue light responses have been reported in algae, fungi, ferns and higher plants.

Some of the typical and most commonly known blue-light responses in plants are:
- Phototropism.
- Stomatal opening.
- Inhibition of hypocotyl elongation.

- Sun tracking by leaves.

- Phototaxis.

- Movements of chloroplasts within the cells.

- Stimulation of synthesis of carotenoids and chlorophylls etc.

Crypto chrome absorbs light rays mostly in violet-blue region of the spectrum (400 – 500 nm). It also absorbs long wave ultraviolet rays in UV-A region (320 to 400 nm). However, most photo-responses of plants caused by crypto chrome result from absorption in violet-blue region of the spectrum but they are simply called as blue-light responses.

Although phytochrome and some other photoreceptors also absorb blue light, but the typical blue-light morphogenetic responses differ from photo-responses mediated by them in being insensitive to red light and there is no red/far- red reversibility.

i.   The action spectra of many blue-light responses in higher plants such as phototropism, stomatal movement, inhibition of hypocotyl elongation etc. are similar and characteristic. They show three peaks in blue region (400 – 500 nm) of the spectrum of visible light. This three peaked, action spectrum is also known as three fingers action spectrum (because of its resemblance in shape with three fingers) and is typical of most blue light responses. Three fingers action spectrum is not observed in phytochrome mediated photo-responses or photo-responses mediated by other photoreceptors other than crypto-chrome.

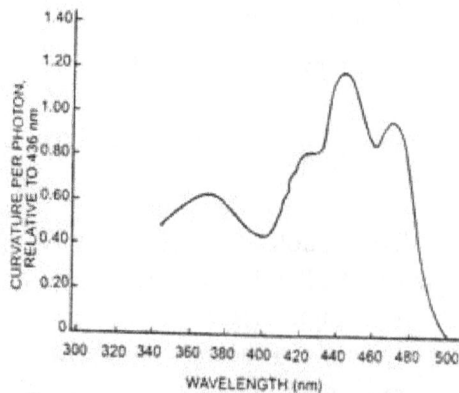

Typical three peaked or three fingers action spectrum for blue light
stimulated phototropism in Avena coleoptile.

ii. Scientists have implicated roles of yellow pigment carotenoids or flavins as photoreceptors in blue-light responses of plants for a long time. However, the spectroscopy of blue-light responses is complex and it is not easy to distinguish between these two types of pigments by comparing available action and absorption spectra.

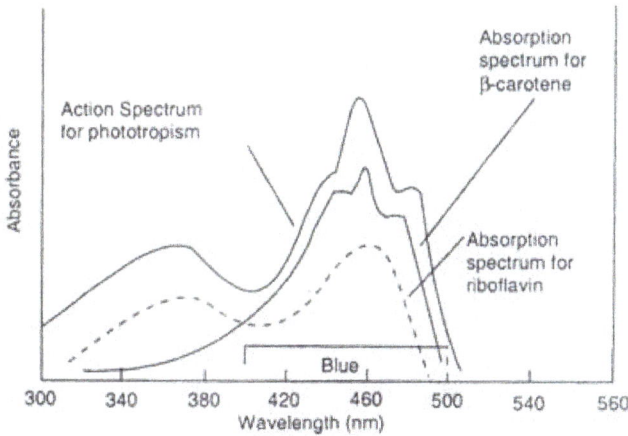

Action spectrum for phototropism with the absorption spectra of riboflavin and carotene

Figure shows relationships between action spectrum for phototropism and absorption spectra of riboflavin and β-carotene. The strong peak in UV-region of the spectrum (360-380 nm) suggests riboflavin as the photoreceptor pigment, while three peaks in blue regions (400 – 500 nm) of the spectrum favours carotene. Nevertheless, accumulating evidences strongly favor flavin pigment to be the primary photoreceptor in phototropism.

Schmidt (1984) has summarized arguments in favor of flavins or carotenoids as photoreceptor pigments in blue light responses of plants as follows:

(a) Arguments in Favor of Flavins

- Action spectra show UV maximum between 350-400 nm.

- Primary steps of the blue light response are dependent on presence of O_2.

- Flavin reactions are often redox reactions.

- Light can be substituted by oxidants while reductants suppress the blue light reaction.

- Blue light reaction is inhibited by flavin inhibitors such as KI.

- Blue light action spectra resemble low temp, spectra of flavins.

- Neurospora mutant which is free of carotenoids shows blue light response.

- Half life of carotenoids in first excited singlet state is very short (10^{-13} seconds)

(b) Arguments in Favor of Carotenoids

- Three peaked (three fingers) action spectra resemble absorption spectra of carotenoids.

- Small or no UV maximum in some action spectra.

- Energy transfer from UV absorbing pigment to carotenoids is feasible.

- Carotenoids from diatom mutant do not show blue light response.

Earlier evidences suggested crypto chrome to be one or both of the yellow pigments, carotenoids (such as β-carotene, zeaxanthin) and/or flavins (such as riboflavin, FAD) which mediate blue-light responses in plants.

However, with extensive researches done with mutants and transgenic plants and over expression studies beginning in early 1990s, the vexed problem of identification of blue-light receptors in plants has gradually been resolved now.

The term crypto chrome is now applied specifically to flavoprotein photoreceptor that mediates inhibition of hypocotyl (stem) elongation caused by blue-light. Blue-light photoreceptor in phototropism and chloroplasts movements in plants is phototropin which is also a flavoprotein. The carotenoid zeaxanthin is blue-light photoreceptor involved in stomatal opening.

Photoreceptors

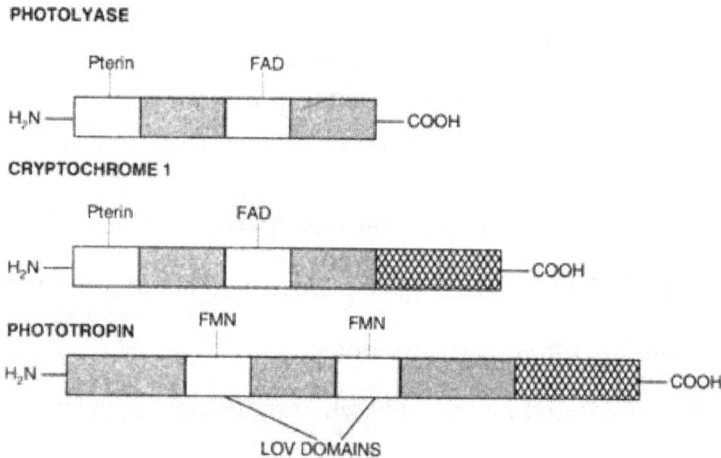

Diagrammatic representation of the chromophore bindin domains of microbial photolyase, cryptochrome 1 (cry 1) and phototropin. Cross-hatched blocks represent carboxy-terminal domains with kinase activity.

Crypto-chrome: The first protein with characteristics of blue-light receptor was isolated in 1993 from Arabidopsis. It was found that hy4 mutant of Arabidopsis had lost the capacity to respond specifically to blue-light in that it showed an elongated hypocotyl even on irradiation with blue-light (In the wild type, blue-light causes inhibition of hypocotyl elongation).

Isolation of the hy4 gene (later named as cry1) showed that it encoded a 75 kDa protein called crypto-chrome 1 (CRY1) with remarkable sequence similarity (homology) to

DNA photolyase in having two chromophores: a flavin adenine dinucleotide (FAD) and a pterin attached to the apoprotein. This led to the establishment of cryptochrome to be a flavoprotein that was involved in inhibition of hypocotyl elongation in response to blue-light. The structure of pterin is given in figure.

(DNA photolyase is a blue-light activated flavoenzyme which repairs UV-induced damage to microbial DNA. Cryptochrome differs from photolyase mainly in two respects. Firstly, the cryptochrome does not show photolyase activity and secondly, unlike photolyase it has an extended carboxy-terminal domain with kinase activity).

A second cryptochrome 2 (CRY2) also with two chromophores like CRY1, has also been isolated from Arabidopsis (Lin 2000). CRY2, mediates blue-light stimulated inhibition of hypocotyl elongation, increase in cotyledon expansion and anthocyanin production. It also has a role in determining flowering time. Both CRY1 and CRY2 appear to be ubiquitous in plant kingdom, but while CRY1 is stable in light grown seedling, CRY2 is rapidly degraded in light.

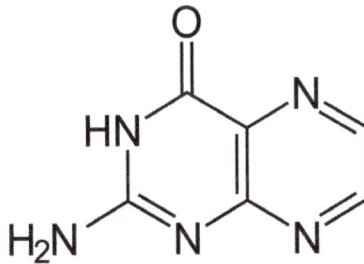

Structure of pterin (fully oxidised)

Mechanism of Action of Cryptochrome

The mechanism of action of crypto chrome remains elusive so far. The flavins are known to participate in oxidation-reduction reactions and photolyases repair damaged DNA (as a result of UV-radiations) by transferring electrons to pyrimidine dimers. Crypto chromes may act probably in a similar way through some electron transfer mechanism.

Phototropins: Phototropins are blue-light receptors that mediate phototropism and chloroplasts movements in plants. In late 1980s, it was found that blue-light stimulated phosphorylation of a 120 kDa protein located on plasma membrane of actively growing regions of etiolated seedlings. These regions were also most responsive to phototropic stimulus. Extensive biochemical and physiological studies showed this protein to be a kinase auto phosphorylating in blue-light and which could be the photoreceptor for phototropism.

Later on, a mutant nph1 (won phototropic hypocotyl 1) was isolated from Arabidopsis which lacked phototropic response in the hypocotyl and also the 120 kDa membrane protein. It was genetically independent of the hy4 mutant as it showed blue-light induced inhibition of hypocotyl elongation.

The nph1 gene was cloned and it was found (as postulated) to encode a 120 kDa protein nph1. The nph1 gene was renamed as phot1 and the protein encoded by it was named phototropin.

Phototropin is also a flavoprotein with two flavin mononucleotide (FMN) chromophores. The protein has a carboxy-terminal domain with a serine/threonine kinase activity. In the amino-terminal half, there are two domains called LOV domains (of about 100 amino acids each) to which are attached the chromophores. (LOV domains are so called because they are characteristics of microbial proteins which regulate response to light, oxygen and voltage).

Recent spectroscopic studies done by Swartz et al, 2000) have shown that in dark, FMN molecules remain non covalently bound to LOV domains, but on irradiations with blue-light they become covalently bound to cysteine residues of the apoprotein through a sulphur atom forming a cysteine- flavin covalent adduct. The reaction is reversed in dark. A second gene called phot 2 has also been isolated from Arabidopsis which is related to phot 1. It is believed that phototropic response involves both phot 1 and phot 2.

Mechanism of Action of Phototropis

The mechanism of action of phototropins is not clear. It has been observed that blue-light causes a transient increase in cytosolic calcium concentration and there are indications that phototropin signalling chain may partly involve regulation of cytoplasmic calcium concentration.

Zeaxanthin: The carotenoid zeaxanthin has been shown to be blue-light receptor in guard cells that plays central role in blue-light stimulated stomatal opening.

Following evidences strongly support role of zeaxanthin in stomatal opening:

- The absorption spectrum of zeaxanthin closely resembles the action spectrum of blue- light stimulated stomatal opening.

- During stomatal opening in intact leaves, the incident radiation, zeaxanthin concentration in guard cell, and stomatal apertures have been found to be directly correlated.

- Blue-light sensitivity of guard cells increases with an increased concentration of zeaxanthin in guard cells.

- There is complete inhibition of blue-light stimulates stomatal opening by 3mM conc. of dithiothreitol (DTT) which is a potent inhibitor of the enzyme that converts violaxanthin to zeaxanthin.

- In facultative CAM plant species such as Mesembryanthemum crystallinum, there is a shift from C_3 to CAM mode of carbon metabolism in response to accumulation of salts. In C_3 mode, the guard cells accumulate zeaxanthin and exhibit blue-light response. But, in CAM mode, neither there is accumulation of zeaxanthin in guard cells nor they respond to blue- light. (In CAM plants, stomata remain closed during the day).

Mechanism of Action of Zeaxanthin

It is believed that the excitation of zeaxanthin by blue- light in guard cells starts a signal transduction pathway that includes:

- Isomerization of zeaxanthin,

- Conformational changes in the apoprotein,

- Transmission of blue-light signal across the chloroplast membrane by a secondary messenger (most probably Ca^{++}, phosphatases, calcium binding protein calmodulin and inositol triphosphate (IP_3),

- Activation of H^+-ATPases at the guard cell plasma membrane resulting in pumping of protons across the membrane and intake of K^+ ions followed by Cl^- ions.

- Turgor builds up in guard cell and stomatal opening.

The blue-light stimulated stomatal opening can be reversed by green light. This may happen if green light is applied with blue-light in continuous light treatment or if a blue-light pulse (of about 30 seconds duration) is followed by a green light pulse. A second blue-light pulse after green-light can restore the stomatal opening. It has been suggested by various workers that green light reverses the isomerization of zeaxanthin resulting in regeneration of inactive zeaxanthin isomer. The latter is unable to mediate the blue-light response.

Besides phytochrome and cryptochrome, there are two other categories of photoreceptors which are known to affect photomorphogenesis in plants. They are:

- Protochlorophyllide-a, a pigment which absorbs red and blue light and is converted to chlorophyll-a, and

- UV-B photoreceptor – one or more unidentified compounds, which absorb short wave ultraviolet rays in UV-B-region (280-320 nm).

ABC Model of Flower Development

Arabidopsis thaliana is an angiosperm belongs to Brassicaceae family. Arabidopsis thaliana possess four concentric whorls or verticils (a circular arrangement of flower which is growing around a central point) which follow an acropetal development and responsible for the formation of sepals, petals, stamen and carpels. As floral meristem determined, its cells will no longer are to divide after get differentiated. ABC model in Arabidopsis contain three classes of homeotic gene called ABC genes which express among four whorls. These three classes of genes specify floral organ identity in the developing flower: Cells of two adjacent floral whorls govern by the activity of each class of genes. Gene belongs to each class is expressed in specific parts of the meristem. yet, there is areas of overlap present in gene expression. Gene, which cause transition of flower meristmatic cell into different parts of the flower, activated by ABC gene. ABC genes product are MADS box contain transcription factors, except AP2, MADS box binds to DNA and a K-box sequence responsible for dimerization. Therefore, these proteins form dimers on DNA. ABC genes activate the expression of other genes that cause transition of flower meristmatic cell into different parts of the flower.

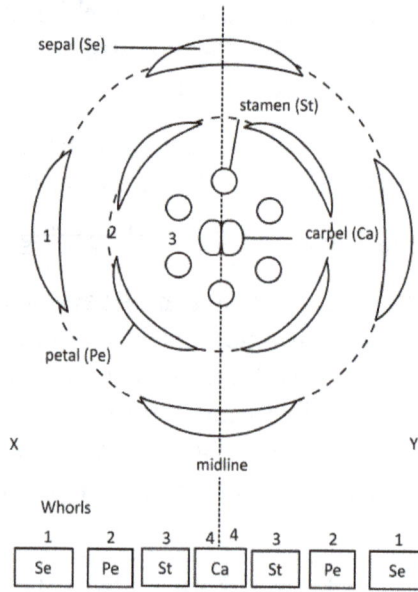

ABC Model of Flower

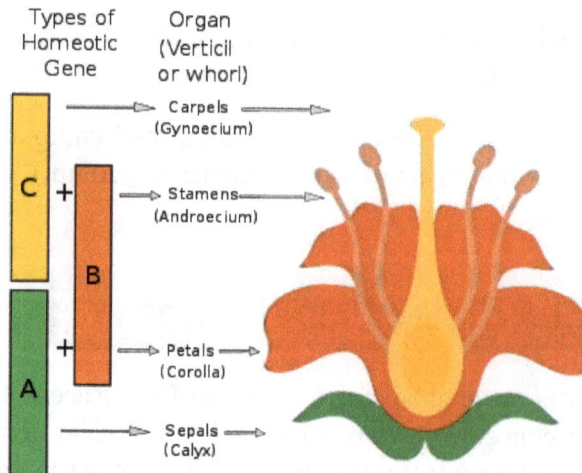

Class A Gene

Class A genes encoded two genes APETALA 1 (AP1) and APETALA 2 (AP2), whose product act as transcription factor and responsible for the formation of sepal and petal within whorls 1 and 2 respectively. Both genes also acts in the floral meristem and AP2 also involve in ovule as well leaves formation.

Class B Gene

Class B genes encoded two genes APETALA 3 (AP3) and PISTILLATA (PI), whose product act as transcription factor and responsible for the formation of petals and stamens within whorls 2 and 3 respectively. Both genes also acts in the floral meristem. Mutation of those gene result in conversion of petals into sepals and of stamens into carpels.

Class C Gene

Class C genes encoded one genes AGAMOUS (AG), whose product act as transcription factor and responsible for the formation of stamens and carpel within whorls 3 and 4 respectively. Genes also acts in the floral meristem. the AG mutants show conversion of androecium or stamen and gynoecium or carpel in to petals and sepals, in their places.

The functional proteins class A inhibit C gene expression in the first two whorls and functional protein of C inhibit class A gene expression in third and fourth whorl. Thus, mutually antagonistic relationship between class A and class C gene, remaining gene get extend due to mutation in class A or class C gene.

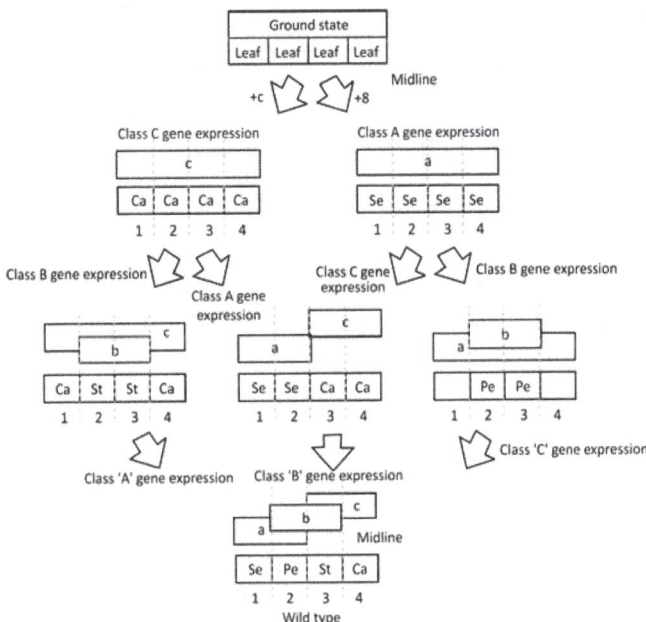

ABC model of flower development

When chemical signal florigen produce by leaves, it is passed to the apical vegetative meristem and turn on the flowering Locus T gene, as a result expression of LEAFY gene takes place. Leafy act as transcription factor for ABC gene and binds to the promoter of ABC genes as dimer, It switch on the ABC gene expression and cause undifferentiated cells in the meristem to change fate and develop as flowers.

A leafy plant result due to mutation in LEAFY because the developmental ground state of a floral organ is a leaf. If there are no ABC transcription factors to switch on floral genes then vegetative meristem carries on making leaves as a result no transition from vegetative growth to flowering

Short terminal flower formation takes place due to over expression of LEAFY. Terminal Flower 1 (TFL1) cause repression of LEAFY activity. A single flower development takes place due to mutation of TFL1 and here LEAFY is no longer repressed. Conversely, flowering is delayed, if TFL1 is constitutively expressed.

Mutational Effect

ABC mutant genes plant produce homeotic mutant flowers.

Class A Mutants

Sepals and petals get transformed into stamens and carpels. This type of mutant called APETALA-1 and APETALA-2. Class A gene does not cause inhibition of Class C genes activity, consequently it get spread into the other whorls.

Class B Mutants

Petals and stamens get transformed into sepals and carpels respectively. This type of mutant called APETALA 3. ABC genes cannot work together in the absence of class B gene, thus petal and stamen development takes place.

Class C Mutants

Stamens and carpels get transformed into petals and sepals. Class C gene does not cause inhibition of Class A gene activity consequently it gets spread into the other whorls. This type of mutant called agamous.

Double Mutant

- Only sepals formation takes place in all four whorls if mutation occur in both class B gene and class C gene.

- Only carpels formation takes place in all four whorls if mutation occurs in both class A gene and class B gene.

- Class A gene and class C gene mutant: leaves formation takes place in whorls first and fourth along with petal and stamen intermediates formation takes place in whorls second and third if mutation occur in both class A gene and class C gene.

References

- Evert, Ray Franklin and Esau, Katherine (2006) Esau's Plant anatomy: meristems, cells, and tissues of the plant body - their structure, function and development Wiley, Hoboken, New Jersey, page xv, ISBN 0-471-73843-3

- Barlow, P (2005). "Patterned cell determination in a plant tissue: The secondary phloem of trees". BioEssays. 27 (5): 533–41. doi:10.1002/bies.20214. PMID 15832381

- Plant-Development: biologyreference.com, Retrieved 19 May 2018

- Jones, Cynthia S. (1999-11-01). "An Essay on Juvenility, Phase Change, and Heteroblasty in Seed Plants". International Journal of Plant Sciences. 160 (S6): –105–S111. doi:10.1086/314215. ISSN 1058-5893. Retrieved 2016-10-16

- Sattler, Rolf (1992). "Process morphology: Structural dynamics in development and evolution". Canadian Journal of Botany. 70 (4): 708–714. doi:10.1139/b92-091

- Megasporogenesis-and-megagametogenesis-in-plants-embryology-biology-77827: biologydiscussion.com, Retrieved 11 March 2018

- Harold C. Bold, C. J. Alexopoulos, and T. Delevoryas. Morphology of Plants and Fungi, 5th ed., page 3. (New York: Harper-Collins, 1987). ISBN 0-06-040839-1

- Sattler, R. (1984). "Homology - a continuing challenge". Systematic Botany. 9 (4): 382–394. doi:10.2307/2418787. JSTOR 2418787

- Importance-of-epigenetic-in-plants-2155-9538-1000151: omicsonline.org, Retrieved 20 May 2018

- Souter, Martin; Lindsey, Keith (June 2000). "Polarity and signaling in plant embryogenesis". Journal of Experimental Botany. 51: 971–983 – via Google Scholar

- The-abc-model-of-flower-development: letstalkacademy.com, Retrieved 31 March 2018

Chapter 3

Reproductive Systems

In flowering plants, the flower is the primary reproductive structure in a plant. It shows great diversity in form and structure and also in methods of reproduction. In flowering and non-flowering plants, there exists a complex interplay between morphological adaptation and influence of environmental factors in sexual reproduction. This chapter has been carefully written to provide an extensive understanding of the varied aspects of sexual and asexual reproduction, and vegetative reproduction in plants.

Plant Reproduction

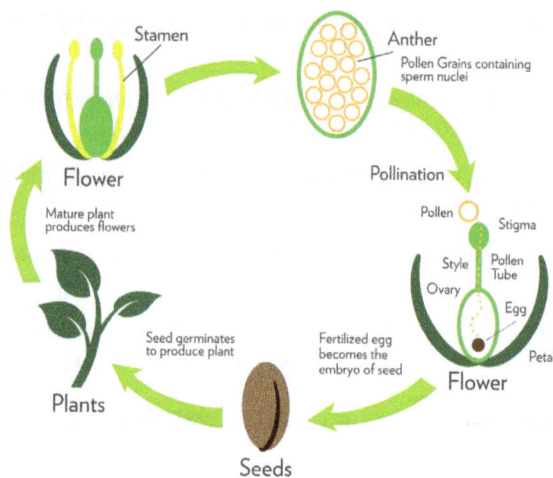

Reproduction (or procreation or breeding) is the biological process by which new individual organisms – "offspring" – are produced from their "parents". Reproduction is a fundamental feature of all known life; each individual organism exists as the result of reproduction. Reproduction is the sine-qua-non of 'Continuity of Life' and without it, all life will cease to exist.

There are two forms of Reproduction: Asexual and Sexual. In asexual reproduction, an organism can reproduce without the involvement of another organism; while Sexual reproduction typically requires the sexual interaction of two specialized organisms, with typically a male fertilizing a female of the same species to create offspring organisms whose genetic characteristics are derived from those of the two parental organisms.

Reproduction in plants is no different. It takes place via both Sexual and Asexual mechanisms and differs according to the plant species. We will learn here in depth about the various modes of reproduction in plants.

Asexual Reproduction in Plants

In asexual reproduction, new plants are produced that are genetically identical clones of the parent plant, and without the contribution of genetic material from another plant. Asexual reproduction in plants can be further divided into two Vegetative Reproduction and Apomixis.

Vegetative Reproduction in Plants

When a a vegetative piece of the original plant such as root, stem or leaf is involved in producing an offspring, it is know as 'Vegetative Reproduction in Plants.' It is often known as a process of 'Survival' and expansion of Biomass.

The broad advantages and disadvantages of Vegetative Reproduction in Plants are listed in the table below:

Advantages of Asexual Vegetative Reproduction in Plants	Disadvantages of Asexual Vegetative Reproduction in Plants
Enables perennial survival of plants	Plants can reproduce only in limited distance
Facilitates expansion in size	Pathogens are transmitted from Parent to Offspring.

The various types of Asexual Vegetative Reproduction in Plants are:

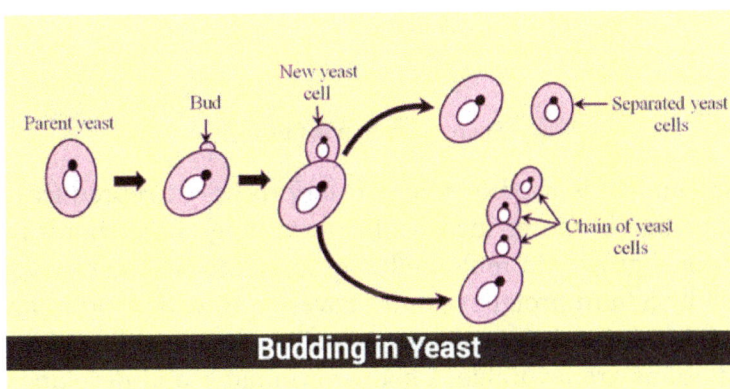

Budding in Yeast

1. Budding

Budding is a form of asexual reproduction in which a new organism develops from an outgrowth or bud due to cell division at one particular site. The new organism remains attached as it grows, separating from the parent organism only when it is mature, leaving behind scar tissue. For example: Yeast is a single-celled organism which repro-duces by this mode.

2. Fragmentation

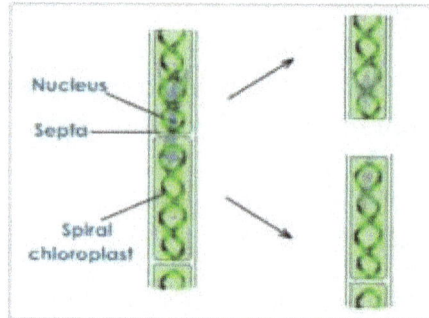

Fragmentation in plants is a form of asexual reproduction or cloning in which an organ-ism is split into fragments. Each of these fragments develop into mature, fully grown individuals that are clones of the original organism. Fragmentation is the mode of re-production in Algae such as Spirogyra.

3. Spore Formation

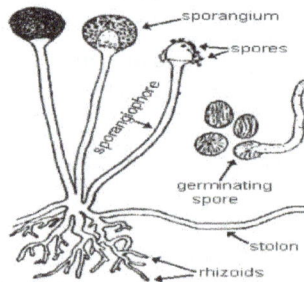

Spore formation in fungus

The term 'Sporogenesis' is used to refer to the process of reproduction in plant via spores. Reproduction via spores involves the spreading of the spores by water or air. Reproductive spores grow into multi-cellular haploid individuals or sporelings. Each spore is covered by a hard protective outer covering to withstand unfavorable condi-tions such as high temperature and low humidity, thus enabling survival for a long time. Once conditions are favorable, a spore germinates and develops into a new in-dividual. The plant organisms which reproduce by this method are Fungi on bread, certain types of moss and ferns, etc.

4. Vegetative Propagation

Vegetative Propagation is a type of asexual reproduction in which new plants are produced from roots, stems, leaves and buds. Since reproduction is through the vegetative parts of the plant, it is known as Vegetative Propagation.

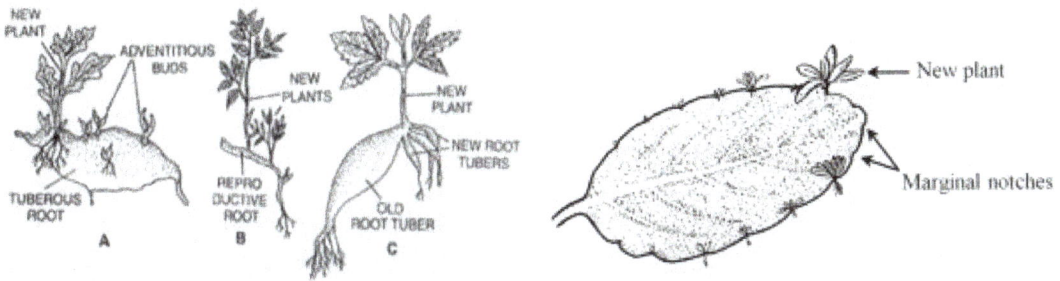

Vegetative propagation by leaves in Bryophylium

Bryophyllum is a plant which reproduces through the vegetative propagation through leaves. Plants such as Sweet Potato and Dahlia reproduce through vegetative propagation by the roots.

- Apomixis

    It is a replacement of the normal sexual reproduction by asexual reproduction without fertilization. Seeds generated by Apomixis are a means of asexual reproduction, involving the formation and dispersal of seeds that do not originate from the fertilization of the embryos. The Offspring is genetically similar to parents. For example: Dandelion.

Formation of Apomictic Seed

- Asexual seed formation from maternal tissues around zygotic embryo inside an ovule.

- Apomictic seed can be formed from nucellar tissues, egg mother cell, flower head (bulbils).

- Found in kentucky bluegrass, dnadelion, citrus, alliums.

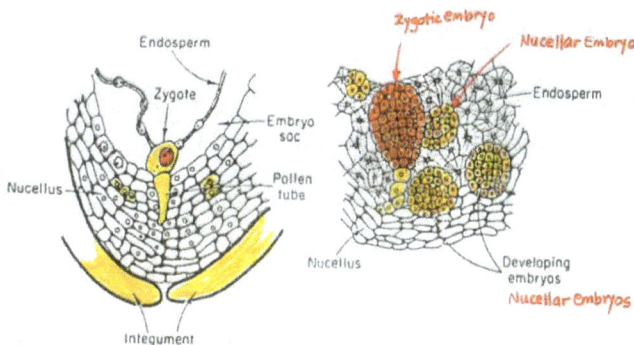

Nucellar embryos in citrus

Sexual Reproduction in Plants

For sexual reproduction of plants, interaction between the male and female species is a prerequisite. The offsprings' genetic structure is not identical but derived from the combination of parent plants.

The sexual reproduction in plants takes place in two phases:

Meiosis

> Through meiosis, the genes of the organism are rearranged and the number of chromosomes is reduced to half i.e., Haploid. The plant produces Gametophytes through the process of meiosis. The Gametophyte then produces the Male or Female Gametes by cell division or mitosis.

Fertilization

> It involves union of both Male and Female 'Gametes' and leads to restoration of the number of chromosomes to form a Diploid Zygote, which then develops into the offspring. The resulting genetic composition has characteristics of both the parent plants from which it is derived.

History of Sexual Reproduction in Plants

- First generation plants were aquatic and released the 'Sperm' or male gametes directly into water; where it would meet with other plants' female counterparts.

- As taller and more complex plants evolved, reproduction was through 'Spores' which would travel along with the wind.

- The seed plants evolved later, which had 'Pollen' a protective covering for the male gamete; which enabled to preserve the viability of the Sperm for greater distances.

Parts of the Reproductive System in Plants

The Flower is The Basic Reproductive Organ In Plants.

A flower is called Unisexual if it has only either the male or the female reproductive parts in it. For example: flowers of Corn, Papaya and Cucumber.

A flower is called bisexual if it has both the male and the female reproductive parts in it. For example: flowers of Mustard, Rose and Petunia.

Structure of the Flower

The flowers are made up of the vegetative parts i.e. the calyx and the corolla and the reproductive parts i.e. the androecium and the gynoecium which are arranged in layers or 'whorls'.

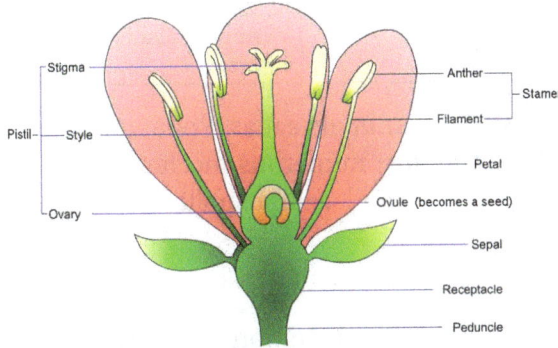

The male reproductive part or the Androecium is made up of units called 'Stamen'. Each Stamen consists of two parts: A stalk called the 'Filament', which is topped by 'Anther.' Pollen Grains' which contain the male gametes or sperms are produced in the Anther.

The female reproductive part or the Gynoecium is the innermost whorl of the flower. It is made up of units called 'Carpel.' Multiple fused carpels form the 'Ovary' where 'Ovules' containing female gametes are produced. Pistil is a structure comprising of fused carpels and has a sticky tip called the Stigma, which acts as a receptor of pollen. The long Stalk acts as a supporting structure and aids development of Pollen tubes from the stigma downwards.

Process of Sexual Reproduction in Plants

The sexual reproduction ins plants is carried out majorly by the process of Pollination. It is a process through which the pollen grain from an anther (male gamete) lands on to the stigma and gradually mates with the ovule (female gamete).

Pollination can be of two types:

Self Pollination	Cross Pollination
Pollen grains are transferred to the stigma of the same flower	Pollen grains are carried to stigma of another flower of the same plant type

Occurs in bisexual plants having anther and stigma maturing at same time	Occurs in unisexual flowers or in bisexual flowers having anther and stigma maturing at different times
Example: Plants like Wheat, Peas	Example: Plants like Lady Finger, Tomato, and Brinjal

For majority of plants, insects or animals act as 'Pollinators' or carriers of pollen from one plant to another. Flowers have numerous mechanisms to attract the pollinators such as color, scent, heat, nectar glands, edible pollen and flower shape.

Process of Fertilization

The formation of a Diploid Zygote by the union of two Haploid male and female gametes is known as 'Fertilization'.

The process begins after the pollen grain sticks to the stigma and begins sending down the pollen tube through which the male gametes pass and unite with the female gametes in the ovary.

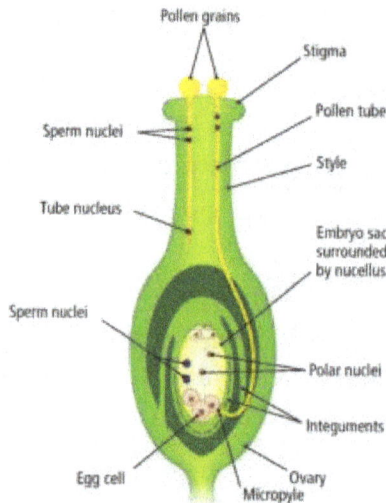

Fertilization in flowering plants is unique among all known organisms, in that not one but two cells are fertilized, in a process called Double Fertilization. One sperm nucleus in the pollen tube fuses with the egg cell in the embryo sac, and the other sperm nucleus fuses with the diploid endosperm nucleus. The fertilized egg cell is a zygote that develops into the diploid embryo of the sporophyte. The fertilized endosperm nucleus develops into the triploid endosperm, a nutritive tissue that sustains the embryo and seedling.

Fruit and Seed Formation

The matured and fertilized ovules form the seeds whereas the fruits are formed from the mature ovaries.

The fruit formed by a single ripened ovary is called as a 'Simple Fruit' For example: Apples, Oranges and Apricots.

The fruit formed by a cluster of matured ovaries, of a single flower is termed 'Aggregated Fruit' for example: Blackberry, Raspberry and Strawberry.

A cluster of many ripened ovaries on separate flowers growing together in the same inflorescence is called a 'Multiple Fruit' for example: Pineapple, Mulberry and Fig.

Vegetative Propagation

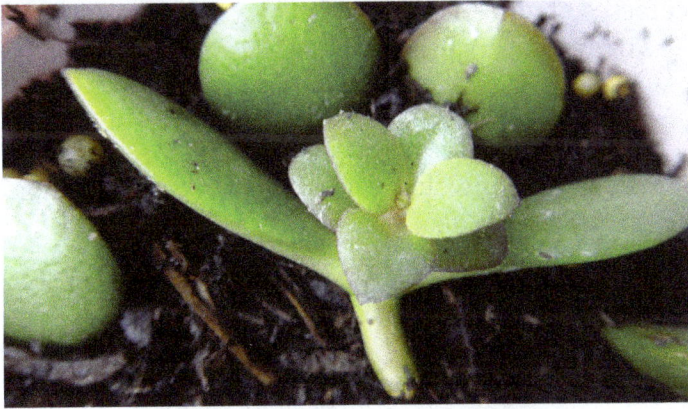

Vegetative propagation is a form of asexual reproduction of a plant. Only one plant is involved and the offspring is the result of one parent. The new plant is genetically identical to the parent.

Vegetative propagation occurs through vegetative plant structures. In non-vascular plants, the vegetative reproductive structures are gemmae and spores whereas, in vascular plants, the roots, stems, leaves, and nodes are the structures that are involved in the propagation. You have learned about the meristem tissue in plants. The same tissue helps in the vegetative propagation. This tissue has undifferentiated cells which divide paving way for the growth of the plant. From the meristems, specialized permanent tissues are formed.

Types of Vegetative Propagation

Vegetative Propagation by Roots

In this process, new plants grow out of the modified roots called tubers. Some plant roots also develop adventitious buds. These buds grow and form new plants/sprouts under the right conditions. These sprouts can be separated from the parent plant and when planted in other areas, new plants are formed. Example – Sweet potato, Dahlia etc.

Vegetative Propagation by Stems

Vegetative propagation occurs through stems when new plants arise from the nodes. This is where buds are formed, which grow into new plants. Stems that grow horizontally on the ground are called runners. As these runners grow, buds are formed at the nodes, which later develop the roots and shoots, resulting in the formation of a new plant. Example – Cyanodon and Mint etc.

The round, swollen part of the underground stem is called a bulb. Within the bulb lies the organ for vegetative propagation such as the central shoot that grows into a new plant. Bulbs have a bud surrounded by layers of fleshy leaves. A few examples include Onions, Garlic, and Tulips etc.

In plants like potatoes, stem tubers are found. This part is the swollen apical part containing many nodes or eyes. Every eye has buds. New plants originate from these buds.

Vegetative Propagation by Leaf

Plants like Bryophyllum, Begonia etc., have adventitious buds coming out from the notches of the leaves. These buds develop into new plants.

Cutting

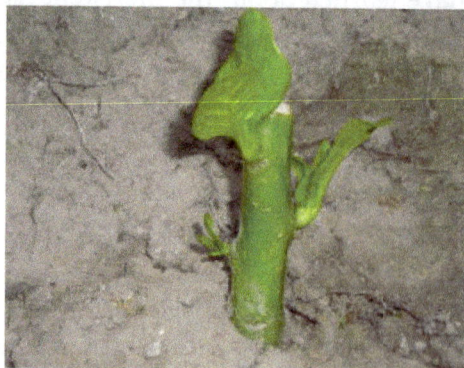

It is the most common method employed by gardeners to grow new plants. A portion of the stem is cut and planted in the soil, which develops roots and further grows into a new plant.

Grafting

In grafting, two closely related plants are used to produce a new plant that has the desired, combined traits of both the parent plants. One plant is the stock, where the root system is taken and the other is the Scion, where the shoot system is used. The scion is attached to the stock of the second plant in this method of artificial vegetative propagation. Grafting is used in a variety of plants like roses, apples, avocado etc.

Budding

In this method, a bud with a small portion of bark is taken from the desired plant. This is inserted into a small slit that is made in the bark of the other plant. Both the plants are tied together and the buds are not allowed to dry.

Asexual Reproduction

Many plants are able to propagate themselves using asexual reproduction. This method does not require the investment required to produce a flower, attract pollinators, or find a means of seed dispersal. Asexual reproduction produces plants that are genetically identical to the parent plant because no mixing of male and female gametes takes place. Traditionally, these plants survive well under stable environmental conditions when compared with plants produced from sexual reproduction because they carry genes identical to those of their parents.

Plants have two main types of asexual reproduction: vegetative reproduction and apomixis. Vegetative reproduction results in new plant individuals without the production of seeds or spores. Many different types of roots exhibit vegetative reproduction. The corm is used by gladiolus and garlic. Bulbs, such as a scaly bulb in lilies and a tunicate bulb in daffodils, are other common examples of this type of reproduction. A potato is a stem tuber, while parsnip propagates from a taproot. Ginger and iris produce rhizomes, while ivy uses an adventitious root (a root arising from a plant part other than the main or primary root), and the strawberry plant has a stolon, which is also called a runner.

Some plants can produce seeds without fertilization. Either the ovule or part of the ovary, which is diploid in nature, gives rise to a new seed. This method of reproduction is known as apomixis.

Roots: Different types of stems allow for asexual reproduction.

(a) The corm of a garlic plant looks similar to (b) a tulip bulb, but the corm is solid tissue, while the bulb consists of layers of modified leaves that surround an underground stem. Both corms and bulbs can self-propagate, giving rise to new plants. (c) Ginger forms masses of stems called rhizomes that can give rise to multiple plants. (d) Potato plants form fleshy stem tubers. Each eye in the stem tuber can give rise to a new plant. (e) Strawberry plants form stolons: stems that grow at the soil surface or just below ground and can give rise to new plants.

An advantage of asexual reproduction is that the resulting plant will reach maturity faster. Since the new plant is arising from an adult plant or plant parts, it will also be sturdier than a seedling. Asexual reproduction can take place by natural or artificial (assisted by humans) means.

Natural and Artificial Methods of Asexual Reproduction in Plants

Plants can undergo natural methods of asexual reproduction, performed by the plant itself, or artificial methods, aided by humans.

Natural Methods of Asexual Reproduction

Natural methods of asexual reproduction include strategies that plants have developed to self-propagate. Many plants, such as ginger, onion, gladioli, and dahlia, continue to grow from buds that are present on the surface of the stem. In some plants, such as the sweet potato, adventitious roots or runners (stolons) can give rise to new plants. In Bryophyllum and kalanchoe, the leaves have small buds on their margins. When these are detached from the plant, they grow into independent plants; they may also start growing into independent plants if the leaf touches the soil. Some plants can be propagated through cuttings alone.

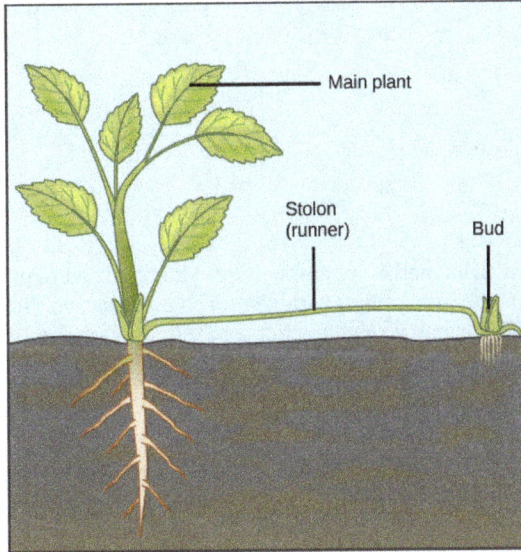

Runners: asexual reproduction: A stolon, or runner, is a stem that runs along the ground. At the nodes, it forms adventitious roots and buds that grow into a new plant.

Artificial Methods of Asexual Reproduction

Artificial methods of asexual reproduction are frequently employed to give rise to new, and sometimes novel, plants. They include grafting, cutting, layering, and micro propagation.

Grafting

Grafting has long been used to produce novel varieties of roses, citrus species, and other plants. In grafting, two plant species are used: part of the stem of the desirable plant is grafted onto a rooted plant called the stock. The part that is grafted or attached is called the scion. Both are cut at an oblique angle (any angle other than a right angle), placed in close contact with each other, and are then held together. Matching up these two surfaces as closely as possible is extremely important because these will be holding the plant together. The vascular systems of the two plants grow and fuse, forming a graft. After a period of time, the scion starts producing shoots, eventually bearing flowers and fruits. Grafting is widely used in viticulture (grape growing) and the citrus industry. Scions capable of producing a particular fruit variety are grafted onto root stock with specific resistance to disease.

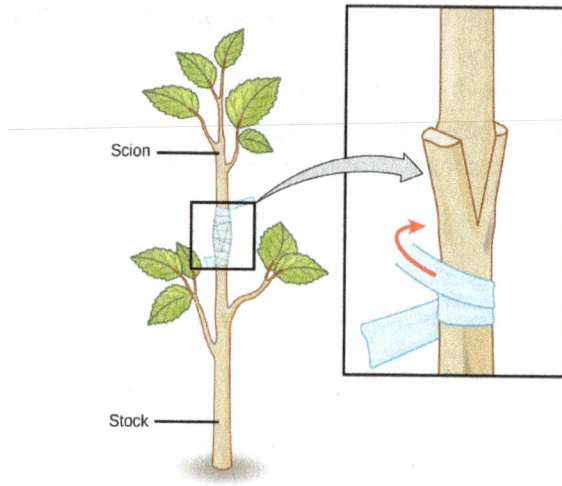

Grafting: Grafting is an artificial method of asexual reproduction used to produce plants combining favorable stem characteristics with favorable root characteristics. The stem of the plant to be grafted is known as the scion, and the root is called the stock.

Cutting

Plants such as coleus and money plant are propagated through stem cuttings where a portion of the stem containing nodes and internodes is placed in moist soil and allowed to root. In some species, stems can start producing a root even when placed only in water. For example, leaves of the African violet will root if kept undisturbed in water for several weeks.

Layering

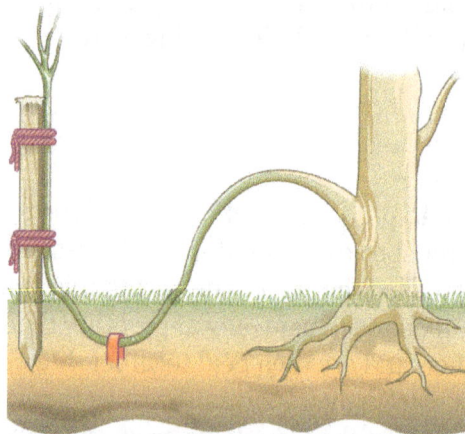

Layering: In layering, a part of the stem is buried so that it forms a new plant.

Layering is a method in which a stem attached to the plant is bent and covered with soil. Young stems that can be bent easily without any injury are the preferred plant for this method. Jasmine and bougainvillea (paper flower) can be propagated this way. In some plants, a modified form of layering known as air layering is employed. A portion of the bark or outermost covering of the stem is removed and covered with moss, which is then taped. Some gardeners also apply rooting hormone. After some time, roots will appear; this portion of the plant can be removed and transplanted into a separate pot.

Micropropagation

Micropropagation (also called plant tissue culture) is a method of propagating a large number of plants from a single plant in a short time under laboratory conditions. This method allows propagation of rare, endangered species that may be difficult to grow under natural conditions, are economically important, or are in demand as disease-free plants.

To start plant tissue culture, a part of the plant such as a stem, leaf, embryo, anther, or seed can be used. The plant material is thoroughly sterilized using a combination of chemical treatments standardized for that species. Under sterile conditions, the plant material is placed on a plant tissue culture medium that contains all the minerals, vitamins, and hormones required by the plant. The plant part often gives rise to an undifferentiated mass, known as a callus, from which, after a period of time, individual plantlets begin to grow. These can be separated; they are first grown under greenhouse conditions before they are moved to field conditions.

Plant Life Spans

The life cycles and life spans of plants vary and are affected by environmental and genetic factors.

Plant life spans: The bristlecone pine, shown here in the Ancient Bristlecone Pine Forest in the White Mountains of eastern California, has been known to live for 4,500 years.

The length of time from the beginning of development to the death of a plant is called its life span. The life cycle, on the other hand, is the sequence of stages a plant goes through from seed germination to seed production of the mature plant. Some plants, such as annuals, only need a few weeks to grow, produce seeds, and die. Other plants, such as the bristlecone pine, live for thousands of years. Some bristlecone pines have a documented age of 4,500 years. Even as some parts of a plant, such as regions containing meristematic tissue (the area of active plant growth consisting of undifferentiated cells capable of cell division) continue to grow, some parts undergo programmed cell death (apoptosis). The cork found on stems and the water-conducting tissue of the xylem, for example, are composed of dead cells.

Annuals, Biennial and Perennials

Plant species that complete their life cycle in one season are known as annuals, an example of which is Arabidopsis, or mouse-ear cress. Biennials, such as carrots, complete their life cycle in two seasons. In a biennial's first season, the plant has a vegetative phase, whereas in the next season, it completes its reproductive phase. Commercial growers harvest the carrot roots after the first year of growth and do not allow the plants to flower. Perennials, such as the magnolia, complete their life cycle in two years or more.

Monocarpic and Polycarpic Plants

In another classification based on flowering frequency, monocarpic plants flower only once in their lifetime; examples of monocarpic plants include bamboo and yucca. During the vegetative period of their life cycle (which may be as long as 120 years in some bamboo species), these plants may reproduce asexually, accumulating a great deal of food material that will be required during their once-in-a-lifetime flowering and setting of seed after fertilization. Soon after flowering, these plants die. Polycarpic plants form flowers many times during their lifetime. Fruit trees, such as apple and orange trees, are polycarpic; they flower every year. Other polycarpic species, such as perennials, flower several times during their life span, but not each year. By this method, the plant does not require all its nutrients to be channeled towards flowering each year.

Genetics and Environmental Conditions

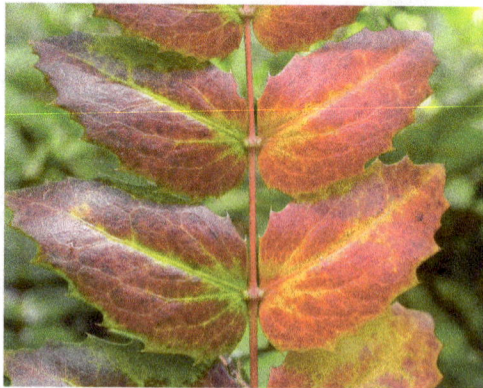

Plant senescence: The autumn color of these Oregon Grape leaves is an example of programmed plant senescence.

As is the case with all living organisms, genetics and environmental conditions have a role to play in determining how long a plant will live. Susceptibility to disease, changing environmental conditions, drought, cold, and competition for nutrients are some of the factors that determine the survival of a plant. Plants continue to grow, despite the presence of dead tissue, such as cork. Individual parts of plants, such as flowers and leaves, have different rates of survival. In many trees, the older leaves turn yellow and eventually fall from the tree. Leaf fall is triggered by factors such as a decrease in photosynthetic efficiency due to shading by upper leaves or oxidative damage incurred as a result of photosynthetic reactions. The components of the part to be shed are recycled by the plant for use in other processes, such as development of seed and storage. This process is known as nutrient recycling. However, the complex pathways of nutrient recycling within a plant are not well understood.

The aging of a plant and all the associated processes is known as senescence, which is marked by several complex biochemical changes. One of the characteristics of senescence is the breakdown of chloroplasts, which is characterized by the yellowing of leaves. The chloroplasts contain components of photosynthetic machinery, such as membranes and proteins. Chloroplasts also contain DNA. The proteins, lipids, and nucleic acids are broken down by specific enzymes into smaller molecules and salvaged by the plant to support the growth of other plant tissues. Hormones are known to play a role in senescence. Applications of cytokinins and ethylene delay or prevent senescence; in contrast, abscissic acid causes premature onset of senescence.

Apomixis

Apomixis is a term commonly used with flowering plants to indicate that they have reproduced asexually through seeds. The plants that grow from these seeds are identical to the mother plant. This is of great use for seed production and plant breeding. Over 400 plant species produce apomictic seeds, including dandelions and blackberries.

Facultative apomixis indicates the situation where apomixis occurs part of the time, but sexual reproduction can still happen. This can be very advantageous evolutionarily. For instance, Kentucky bluegrass periodically produces new strains through sexual reproduction. The best of them are propagated through apomixis, and the plant has many strains that are well-adapted to localized regions.

Arctic plants benefit greatly from apomictic reproduction. The extreme conditions there make pollination by insects difficult, so it is hard to transfer genetic material from plant to plant. Given the difficult conditions, the plants might benefit from having a few more greatly specialized strains than from the constantly evolving populations that would occur with sexual reproduction.

There is a lot of interest in learning how to genetically manipulate apomictic plants for agricultural reasons. There are, however, technical difficulties in doing so. Plants that are apomictic frequently exhibit polyploidy, meaning they have more than two copies of their chromosomes. This can make gene transfer difficult.

An advantage to using apomictic seed would be that small farmers would be able to produce their own seeds of elite cultivars. Viral diseases can spread through vegetatively propagated plants, and this could be minimized by cloning with seed cultures. One could also take advantage of locally-adapted varieties that are resistant to extreme climatic conditions or pathogens.

There are two main forms of apomixis. The first involves the gametophyte, which is the plant's reproductive haploid multicellular structure. Being haploid means that it only contains one set of chromosomes. The gametophyte produces gametes — mature reproductive cells that normally would unite with another of the opposite sex to undergo sexual reproduction. In gametophytic apomixis, however, an unfertilized egg cell gives rise to an embryo. This is similar to parthenogenesis in animals, which is reproduction in females, without sex.

The second form is known as sporophytic apomixis. In this case, the embryos are formed like buds directly from integument tissue. This is part of the inner tissue of the ovule, the structure that holds the embryo sac. The embryo sac continues with its development, while the sporophytic embryo will go ahead to form apomictic seeds. This process is also known as adventitious embryony. Except for commonly occurring in citrus plants, sporophytic apomixis is a rare process in higher plants.

Apogamy and Apospory in Non-flowering Plants

The gametophytes of bryophytes, and less commonly ferns and lycopods can develop a group of cells that grow to look like a sporophyte of the species but with the ploidy level of the gametophyte, a phenomenon known as apogamy. The sporophytes of plants of these groups may also have the ability to form a plant that looks like a gametophyte but with the ploidy level of the sporophyte, a phenomenon known as apospory.

In Flowering Plants (Angiosperms)

Agamospermy, asexual reproduction through seeds, occurs in flowering plants through many different mechanisms and a simple hierarchical classification of the different types is not possible. Consequently, there are almost as many different usages of terminology for apomixis in angiosperms as there are authors on the subject. For English speakers, Maheshwari 1950 is very influential. German speakers might prefer to consult Rutishauser 1967. Some older text books on the basis of misinformation (that the egg cell in a meiotically unreduced gametophyte can never be fertilized) attempted to reform the terminology to match the term parthenogenesis as it is used in zoology, and this continues to cause much confusion.

Agamospermy occurs mainly in two forms: In gametophytic apomixis, the embryo arises from an unfertilized egg cell (i.e. by parthenogenesis) in a gametophyte that was produced from a cell that did not complete meiosis. In adventitious embryony (sporophytic apomixis), an embryo is formed directly (not from a gametophyte) from nucellus or integument tissue.

Types in Flowering Plants

Caribbean Agave producing plantlets on the old flower stem.

Maheshwari used the following simple classification of types of apomixis in flowering plants:

- Nonrecurrent apomixis: In this type "the megaspore mother cell undergoes the usual meiotic divisions and a haploid embryo sac [megagametophyte] is formed. The new embryo may then arise either from the egg (haploid parthenogenesis) or from some other cell of the gametophyte (haploid apogamy)." The haploid plants have half as many chromosomes as the mother plant, and "the process is not repeated from one generation to another" (which is why it is called nonrecurrent).

- Recurrent apomixis, is now more often called gametophytic apomixis: In this type, the megagametophyte has the same number of chromosomes as the mother plant because meiosis was not completed. It generally arises either from an archesporial cell or from some other part of the nucellus.

- Adventive embryony, also called sporophytic apomixis, sporophytic budding, or nucellar embryony: Here there may be a megagametophyte in the ovule, but the embryos do not arise from the cells of the gametophyte; they arise from cells of nucellus or the integument. Adventive embryony is important in several species of Citrus, in Garcinia, Euphorbia dulcis, Mangifera indica etc.

- Vegetative apomixis: In this type "the flowers are replaced by bulbils or other vegetative propagules which frequently germinate while still on the plant".

Vegetative apomixis is important in Allium, Fragaria, Agave, and some grasses, among others.

Types of Gametophytic Apomixis

Gametophytic apomixis in flowering plants develops in several different ways. A megagametophyte develops with an egg cell within it that develops into an embryo through parthenogenesis. The central cell of the megagametophyte may require fertilization to form the endosperm, pseudogamous gametophytic apomixis, or in autonomous gametophytic apomixis fertilization is not required.

- In diplospory (also called generative apospory), the megagametophyte arises from a cell of the archesporium.

- In apospory (also called somatic apospory), the megagametophyte arises from some other nucellus cell.

Considerable confusion has resulted because diplospory is often defined to involve the megaspore mother cell only, but a number of plant families have a multicellular archesporium and the megagametophyte could originate from another archesporium cell.

Diplospory is further subdivided according to how the megagametophyte forms:

- Allium odorum–A nutans type: P The chromosomes double (endomitosis) and then meiosis proceeds in an unusual way, with the chromosome copies pairing up (rather than the original maternal and paternal copies pairing up).

- Taraxacum type: Meiosis I fails to complete, meiosis II creates two cells, one of which degenerates; three mitotic divisions form the megagametophyte.

- Ixeris type: Meiosis I fails to complete; three rounds of nuclear division occur without cell-wall formation; wall formation then occurs.

- Blumea–Elymus types: A mitotic division is followed by degeneration of one cell; three mitotic divisions form the megagametophyte.

- Antennaria–Hieracium types: three mitotic divisions form the megagametophyte.

- Eragrostis–Panicum types: Two mitotic division give a 4-nucleate megagametophyte, with cell walls to form either three or four cells.

Incidence in Flowering Plants

Apomixis occurs in at least 33 families of flowering plants, and has evolved multiple times from sexual relatives. Apomictic species or individual plants often have a hybrid origin, and are usually polyploid.

In plants with both apomictic and meiotic embryology, the proportion of the different types can differ at different times of year, and photoperiod can also change the proportion. It appears unlikely that there are any truly completely apomictic plants, as low rates of sexual reproduction have been found in several species that were previously thought to be entirely apomictic.

The genetic control of apomixis can involve a single genetic change that affects all the major developmental components, formation of the megagametophyte, parthenogenesis of the egg cell, and endosperm development. However, the timing of the various developmental processes is critical to successful development of an apomictic seed, and the timing can be affected by multiple genetic factors.

Parthenocarpy

The condition in which fruits are developed without the formation of seeds. This is mainly due to the absence of fertilization, pollination and embryo development. In botany, parthenocarpy means virgin fruit. These types of fruits are generally seedless. This process of fruits was introduced in the year 1902.

During cultivation, parthenocarpy is introduced along with other plant hormones including gibberellin acid and it results in maturing of the ovaries without the process of fertilization and produces bigger and pulpy fruits. This process is applicable to all kinds of crops from varieties of squash to cucumber and lot more.

Banana is a good example of parthenocarpy. In this natural process, the produced bananas are sterile, developed without the viable ovaries and do not produce seeds, which mean they must propagate vegetative. Pineapples and figs are also examples of parthenocarpy which occur naturally.

PARTHENOCARPY

Types of Parthenocarpy

Parthenocarpy can be categorized into two parts, which are:

- Vegetative parthenocarpy – This generally takes place without pollination and due to the absence of pollination, no seeds are produced within the fruits.

- Stimulative parthenocarpy – This generally takes place without the process of fertilization. This condition occurs when the ovipositor of a wasp is inserted into the ovary of a flower. This can also be achieved by flowing air or growth regulators into the unisexual flowers that are present inside the syconium.

Process of Parthenocarpy

Vegetative parthenocarpy in plants, like pear and fig, take place without pollination. As we know, pollination leads to fertilization, so in the absence of pollination, no seeds can form.

Stimulative parthenocarpy is a process where pollination is required but no fertilization takes place. It occurs when a wasp inserts its ovipositor into the ovary of a flower. It can also be simulated by blowing air or growth hormones into the unisexual flowers found inside something called a syconium. The syconium is basically the flask-shaped structure lined with the unisexual flowers.

Growth regulating hormones, when used on crops, also halt the fertilization process. In some crop plants, this also occurs due to genome manipulation.

Benefits of Parthenocarpy

- This is more healthful and easier to achieve results.

- Provides seedless fruits and improves the quality.

- It reduces the complete cost of the cultivation.

- This improves crop yield even without using organic pesticides.

- Plant growth regulators are natural and the fruits produced are larger.

- Parthenocarpy helps the grower to keep the insects and pests away without even using chemicals. Because there is no requirement of pollinating insects for the formation of fruits. So the plants can be covered from getting attacked by the bad insects.

Apogamy

Apogamy is the formation of a sporophyte from vegetative cells of a gametophyte. It can be divided into two types, obligate and induced. Obligate apogamy is the normal means of reproduction in certain fern species. It has been estimated that 12 percent of the cytologically investigated homosporous ferns are apogamous and in the other i.e. normally sexually reproducing species apogamy can be induced in response to specific cultural condition. The induction and promotion of apogamy in vitro in ferns has been investigated by Whittier and co-workers with an emphasis on factors such as carbohydrates, ethylene and osmotica. The tern apogamy is now applied to the process ofsporophyte production from callus induced or gametophytes. Such apogamy has been reported in several ferns e.g. Platycerium coronarium, Dryopteris affinis, and Osmunda regalis.

Apospory

Apospoiy is the process of development of gametophytes from the vegetative parts ofthe sporophyte without the intervention of spores. Apospoiy was noticed in many ferns and some workers have induced it under culture condition e.g. White, Hirsch, Sheffield and Bell, Raghavan, Materi and Cumming, Sheffield, Camloh, Fernandez and Dolinsek and Camloh. Ragjhavan, proposed die following explanation of apospory induction. The genetic blue-print of the sporophyte cells is activated under unfavourable conditions to produce structures that are nutritionally less demanding than die sporophyte itself, in order to prolong the life ofthe plant to the genetically permissible extent.

It is of two types:

(i)   Generative or haploid apospory: If the embryo sac develops from one of the megaspores (n), the process is called generative or haploid apospory. Since it cannot regenerate, as it is haploid and fertilization fails, the process gives rise to non-recurrent apomicts.

(ii) Somatic or diploid apospory: When diploid embryo sac is formed from nucellus or other cells, the process is termed as somatic or diploid apospory. Since it regenerates without fertilization, it is recurrent.

Sexual Reproduction

Reproduction in plants takes place sexually and asexually as well. But the majority of the flowering plants reproduce sexually. The flower is the reproductive part of a plant i.e., both male and female gametes are produced by flowers. Sexual reproduction in plants takes place in flowers.

Parts of a Flower in Plant Reproduction

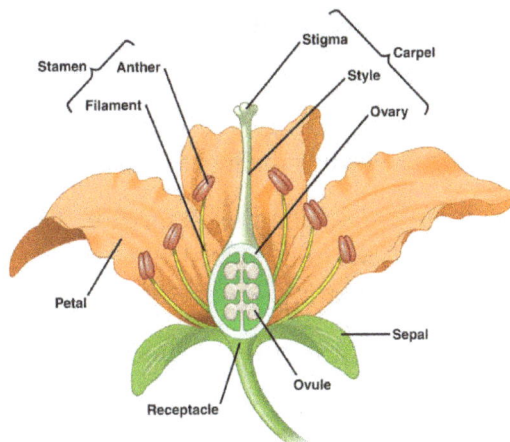

A bisexual flower typically contains the male and female parts in it. There are other supporting structures as well apart from the reproductive parts for sexual reproduction.

There are four main layers of the parts of a flower:

- Calyx

- Corolla

- Androecium

- Gynoecium

Calyx

It is a collection of sepals. The sepals are the green colored small florets that are considered the first layers of the flower from the base. In some cases, the sepals have color. They are called petaloid. Their main function is to protect the flower while it is still in the bud stage.

Corolla

This layer is a collection of petals. It is the second layer of the flower, superior to the calyx layer. The petals are the colorful part of a flower that helps to attract insects and birds to the flower to facilitate pollination.

Androecium

It is the third layer of the flower superior to the Corolla. This is a term given to the male parts for sexual reproduction of a plant. The androecium is made up of a collection of stamens. Each stamen has the following parts:

- Anther- It is present at the tip of the filament. It is internally lobed. Pollen grains are inside the Anther Lobe.

- Filament- It is a thin stalk-like structure that holds the anther

Gynoecium

A Gynoecium is a collection of carpels. It is the fourth layer of a flower. It has three parts:

- Stigma- It is a small and sticky landing structure. The pollen grain from the same or different flower sticks to it. This structure acts as a landing for the insects or birds that act as pollinating agents.

- Style- It is a thin stalk-like structure that holds the stigma.

- Ovary- It is the base of the style and contains the ovules which contain the female gametes.

Pollination and Fertilization in Plant Reproduction

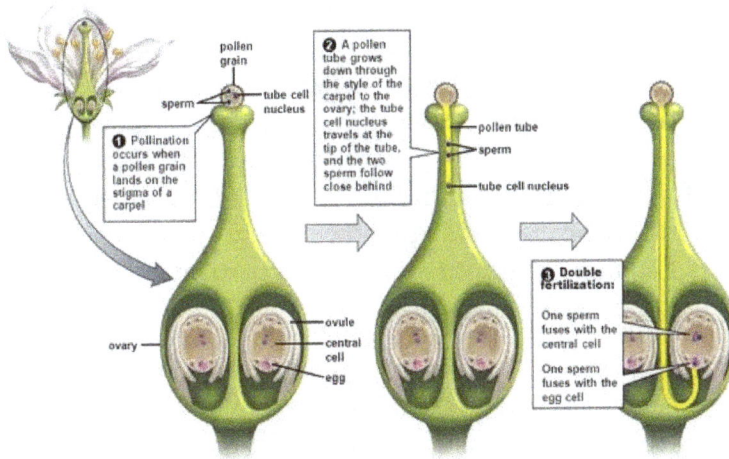

Pollination and Fertilization of a flower

The transfer of pollen grains from the anther of one flower to the stigma of the same or another flower is known as pollination. It can be caused by insects, birds, wind, water and animals including man. These are together called as pollinating agents.

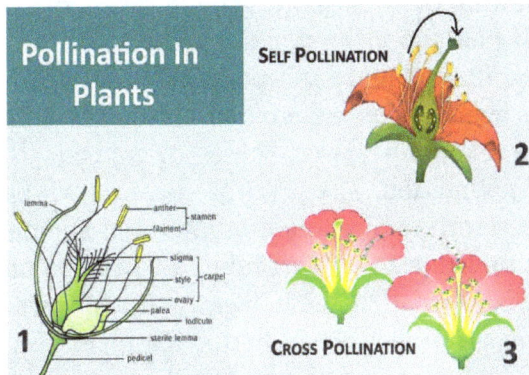

Self-Pollination and Cross-Pollination

In angiosperms, pollination is defined as the placement or transfer of pollen from the anther to the stigma of the same flower or another flower. In gymnosperms, pollination involves pollen transfer from the male cone to the female cone. Upon transfer, the pollen germinates to form the pollen tube and the sperm for fertilizing the egg. Pollination has been well studied since the time of Gregor Mendel. Mendel successfully carried out self- as well as cross-pollination in garden peas while studying how characteristics were passed on from one generation to the next. Today's crops are a result of plant breeding, which employs artificial selection to produce the present-day cultivars. A case in point is today's corn, which is a result of years of breeding that started with its ancestor, teosinte. The teosinte that the ancient Mayans originally began cultivating had tiny seeds—vastly different from today's relatively giant ears of corn. Interestingly, though these two plants appear to be entirely different, the genetic difference between them is miniscule.

Pollination takes two forms: self-pollination and cross-pollination. Self-pollination occurs when the pollen from the anther is deposited on the stigma of the same flower, or another flower on the same plant. Cross-pollination is the transfer of pollen from the anther of one flower to the stigma of another flower on a different individual of the same species. Self-pollination occurs in flowers where the stamen and carpel mature at the same time, and are positioned so that the pollen can land on the flower's stigma. This method of pollination does not require an investment from the plant to provide nectar and pollen as food for pollinators.

Living species are designed to ensure survival of their progeny; those that fail become extinct. Genetic diversity is therefore required so that in changing environmental or stress conditions, some of the progeny can survive. Self-pollination leads to the production of plants with less genetic diversity, since genetic material from the same plant is used to form gametes, and eventually, the zygote. In contrast, cross-pollination—or out-crossing—leads to greater genetic diversity because the microgametophyte and megagametophyte are derived from different plants.

Because cross-pollination allows for more genetic diversity, plants have developed many ways to avoid self-pollination. In some species, the pollen and the ovary mature at different times. These flowers make self-pollination nearly impossible. By the time pollen matures and has been shed, the stigma of this flower is mature and can only be pollinated by pollen from another flower. Some flowers have developed physical features that prevent self-pollination. The primrose is one such flower. Primroses have evolved two flower types with differences in anther and stigma length: the pin-eyed flower has anthers positioned at the pollen tube's halfway point, and the thrum-eyed flower's stigma is likewise located at the halfway point. Insects easily cross-pollinate while seeking the nectar at the bottom of the pollen tube. This phenomenon is also known as heterostyly. Many plants, such as cucumber, have male and female flowers located on different parts of the plant, thus making self-pollination difficult. In yet other species, the male and female flowers are borne on different plants (dioecious). All of these are barriers to self-pollination; therefore, the plants depend on pollinators to transfer pollen. The majority of pollinators are biotic agents such as insects (like bees, flies, and butterflies), bats, birds, and other animals. Other plant species are pollinated by abiotic agents, such as wind and water.

Methods of Pollination

Pollination by Insects

Bees are perhaps the most important pollinator of many garden plants and most commercial fruit trees. The most common species of bees are bumblebees and honeybees.

Since bees cannot see the color red, bee-pollinated flowers usually have shades of blue, yellow, or other colors. Bees collect energy-rich pollen or nectar for their survival and energy needs. They visit flowers that are open during the day, are brightly colored, have a strong aroma or scent, and have a tubular shape, typically with the presence of a nectar guide. A nectar guide includes regions on the flower petals that are visible only to bees, and not to humans; it helps to guide bees to the center of the flower, thus making the pollination process more efficient. The pollen sticks to the bees' fuzzy hair, and when the bee visits another flower, some of the pollen is transferred to the second flower. Recently, there have been many reports about the declining population of honeybees. Many flowers will remain unpollinated and not bear seed if honeybees disappear. The impact on commercial fruit growers could be devastating.

Figure: Insects, such as bees, are important agents of pollination.

Many flies are attracted to flowers that have a decaying smell or an odor of rotting flesh. These flowers, which produce nectar, usually have dull colors, such as brown or purple. They are found on the corpse flower or voodoo lily (Amorphophallus), dragon arum (Dracunculus), and carrion flower (Stapleia, Rafflesia). The nectar provides energy, whereas the pollen provides protein. Wasps are also important insect pollinators, and pollinate many species of figs.

Figure: A corn earworm sips nectar from a night-blooming Gaura plant.

Butterflies, such as the monarch, pollinate many garden flowers and wildflowers, which usually occur in clusters. These flowers are brightly colored, have a strong fragrance, are open during the day, and have nectar guides to make access to nectar easier. The pollen is picked up and carried on the butterfly's limbs. Moths, on the other hand, pollinate flowers during the late afternoon and night. The flowers pollinated by moths are pale or white and are flat, enabling the moths to land. One well-studied example of a moth-pollinated plant is the yucca plant, which is pollinated by the yucca moth. The shape of the flower and moth have adapted in such a way as to allow successful pollination. The moth deposits pollen on the sticky stigma for fertilization to occur later. The female moth also deposits eggs into the ovary. As the eggs develop into larvae, they obtain food from the flower and developing seeds. Thus, both the insect and flower benefit from each other in this symbiotic relationship. The corn earworm moth and Gaura plant have a similar relationship.

Pollination by Bats

In the tropics and deserts, bats are often the pollinators of nocturnal flowers such as agave, guava, and morning glory. The flowers are usually large and white or pale-colored; thus, they can be distinguished from the dark surroundings at night. The flowers have a strong, fruity, or musky fragrance and produce large amounts of nectar. They are naturally large and wide-mouthed to accommodate the head of the bat. As the bats seek the nectar, their faces and heads become covered with pollen, which is then transferred to the next flower.

Pollination by Birds

Figure: Hummingbirds have adaptations that allow them to reach the nectar of certain tubular flowers.

Many species of small birds, such as the hummingbird and sun birds, are pollinators for plants such as orchids and other wildflowers. Flowers visited by birds are usually sturdy and are oriented in such a way as to allow the birds to stay near the flower without getting their wings entangled in the nearby flowers. The flower typically has a curved, tubular shape, which allows access for the bird's beak. Brightly colored, odorless flowers that are open during the day are pollinated by birds. As a bird seeks energy-rich nectar, pollen is deposited on the bird's head and neck and is then transferred to the next flower it visits. Botanists have been known to determine the range of extinct plants by collecting and identifying pollen from 200-year-old bird specimens from the same site.

Pollination by Wind

Most species of conifers, and many angiosperms, such as grasses, maples and oaks, are pollinated by wind. Pine cones are brown and unscented, while the flowers of wind-pollinated angiosperm species are usually green, small, may have small or no petals, and produce large amounts of pollen. Unlike the typical insect-pollinated flowers, flowers adapted to pollination by wind do not produce nectar or scent. In wind-pollinated species, the microsporangia hang out of the flower, and, as the wind blows, the lightweight pollen is carried with it.

Figure: A person knocks pollen from a pine tree.

The flowers usually emerge early in the spring, before the leaves, so that the leaves do not block the movement of the wind. The pollen is deposited on the exposed feathery stigma of the flower.

(a)   (b)

Figure: These male (a) and female (b) catkins are from the goat willow tree (Salix caprea).
Note how both structures are light and feathery to better disperse and catch the wind-blown pollen.

Pollination by Water

Some weeds, such as Australian sea grass and pond weeds, are pollinated by water. The pollen floats on water, and when it comes into contact with the flower, it is deposited inside the flower.

Pollination by Deception

Orchids are highly valued flowers, with many rare varieties. They grow in a range of specific habitats, mainly in the tropics of Asia, South America, and Central America. At least 25,000 species of orchids have been identified.

Flowers often attract pollinators with food rewards, in the form of nectar. However, some species of orchid are an exception to this standard: they have evolved different ways to attract the desired pollinators. They use a method known as food deception, in which bright colors and perfumes are offered, but no food. Anacamptis morio, commonly known as the green-winged orchid, bears bright purple flowers and emits a strong scent. The bumblebee, its main pollinator, is attracted to the flower because of the strong scent—which usually indicates food for a bee—and in the process, picks up the pollen to be transported to another flower.

Figure: Certain orchids use food deception or sexual deception to attract pollinators. Shown here is a bee orchid (Ophrys apifera).

Other orchids use sexual deception. Chiloglottis trapeziformis emits a compound that smells the same as the pheromone emitted by a female wasp to attract male wasps. The male wasp is attracted to the scent, lands on the orchid flower, and in the process, transfers pollen. Some orchids, like the Australian hammer orchid, use scent as well as visual trickery in yet another sexual deception strategy to attract wasps. The flower of this orchid mimics the appearance of a female wasp and emits a pheromone. The male wasp tries to mate with what appears to be a female wasp, and in the process, picks up pollen, which it then transfers to the next counterfeit mate.

Double Fertilization

After pollen is deposited on the stigma, it must germinate and grow through the style to reach the ovule. The microspores, or the pollen, contain two cells: the pollen tube cell and the generative cell. The pollen tube cell grows into a pollen tube through which the generative cell travels. The germination of the pollen tube requires water, oxygen, and certain chemical signals. As it travels through the style to reach the embryo sac, the pollen tube's growth is supported by the tissues of the style. In the meantime, if the generative cell has not already split into two cells, it now divides to form two sperm cells. The pollen tube is guided by the chemicals secreted by the synergids present in the embryo sac, and it enters the ovule sac through the micropyle. Of the two sperm cells, one sperm fertilizes the egg cell, forming a diploid zygote; the other sperm fuses with the two polar nuclei, forming a triploid cell that develops into the endosperm. Together, these two fertilization events in angiosperms are known as double fertilization. After fertilization is complete, no other sperm can enter. The fertilized ovule forms the seed, whereas the tissues of the ovary become the fruit, usually enveloping the seed.

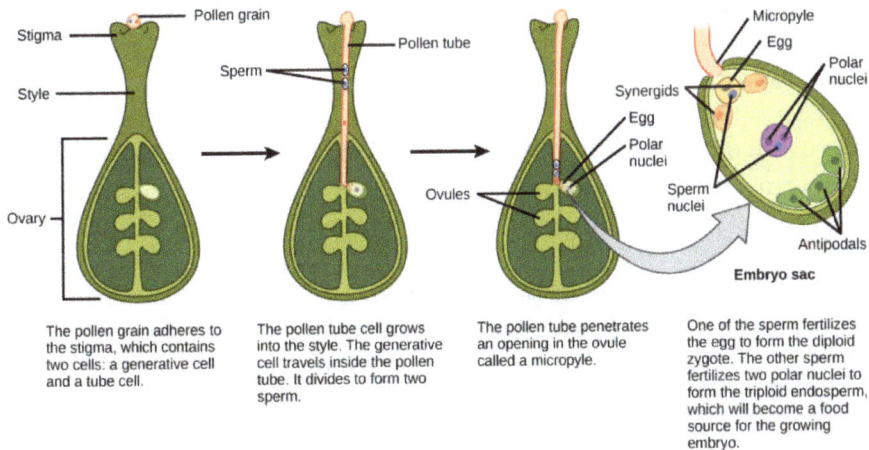

| The pollen grain adheres to the stigma, which contains two cells: a generative cell and a tube cell. | The pollen tube cell grows into the style. The generative cell travels inside the pollen tube. It divides to form two sperm. | The pollen tube penetrates an opening in the ovule called a micropyle. | One of the sperm fertilizes the egg to form the diploid zygote. The other sperm fertilizes two polar nuclei to form the triploid endosperm, which will become a food source for the growing embryo. |

Figure: In angiosperms, one sperm fertilizes the egg to form the 2n zygote, and the other sperm fertilizes the central cell to form the 3n endosperm. This is called a double fertilization.

After fertilization, the zygote divides to form two cells: the upper cell, or terminal cell, and the lower, or basal, cell. The division of the basal cell gives rise to the suspensor, which eventually makes connection with the maternal tissue. The suspensor provides a route for nutrition to be transported from the mother plant to the growing embryo. The terminal cell also divides, giving rise to a globular-shaped proembryo. In dicots (eudicots), the developing embryo has a heart shape, due to the presence of the two rudimentary cotyledons. In non-endospermic dicots, such as Capsella bursa, the endosperm develops initially, but is then digested, and the food reserves are moved into the two cotyledons. As the embryo and cotyledons enlarge, they run out of room inside the developing seed, and are forced to bend. Ultimately, the embryo and cotyledons fill the seed, and the seed is ready for dispersal. Embryonic development is suspended after some time, and growth is resumed only when the seed germinates. The developing seedling will rely on the food reserves stored in the cotyledons until the first set of leaves begin photosynthesis.

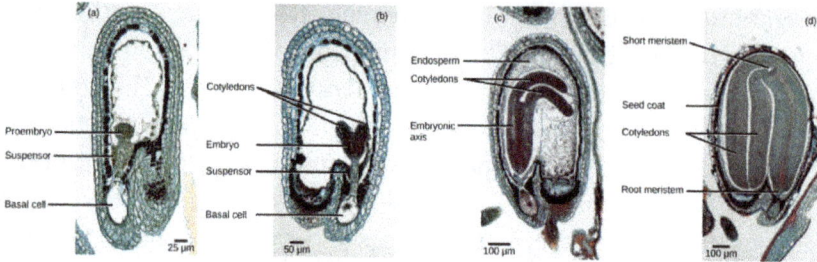

Figure: Shown are the stages of embryo development in the ovule of a shepherd's purse (Capsella bursa). After fertilization, the zygote divides to form an upper terminal cell and a lower basal cell.

(a) In the first stage of development, the terminal cell divides, forming a globular pro-embryo. The basal cell also divides, giving rise to the suspensor. (b) In the second stage, the developing embryo has a heart shape due to the presence of cotyledons. (c) In the third stage, the growing embryo runs out of room and starts to bend. (d) Eventually, it completely fills the seed.

Development of a Seed

The mature ovule develops into the seed. A typical seed contains a seed coat, cotyledons, endosperm, and a single embryo.

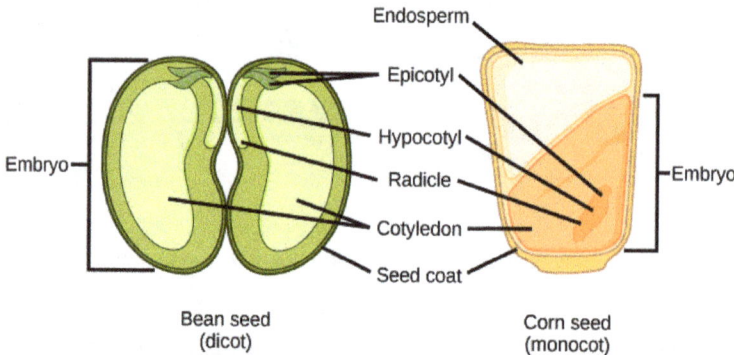

Figure: The structures of dicot and monocot seeds are shown.

Dicots (left) have two cotyledons. Monocots, such as corn (right), have one cotyledon, called the scutellum; it channels nutrition to the growing embryo. Both monocot and dicot embryos have a plumule that forms the leaves, a hypocotyl that forms the stem, and a radicle that forms the root. The embryonic axis comprises everything between the plumule and the radicle, not including the cotyledon(s).

The storage of food reserves in angiosperm seeds differs between monocots and dicots. In monocots, such as corn and wheat, the single cotyledon is called a scutellum; the scutellum is connected directly to the embryo via vascular tissue (xylem and phloem). Food reserves are stored in the large endosperm. Upon germination, enzymes are secreted by the aleurone, a single layer of cells just inside the seed coat that surrounds the endosperm and embryo. The enzymes degrade the stored carbohydrates, proteins and lipids, the products of which are absorbed by the scutellum and transported via a vasculature strand to the developing embryo. Therefore, the scutellum can be seen to be an absorptive organ, not a storage organ.

The two cotyledons in the dicot seed also have vascular connections to the embryo. In endospermic dicots, the food reserves are stored in the endosperm. During germination, the two cotyledons therefore act as absorptive organs to take up the enzymatically released food reserves, much like in monocots (monocots, by definition, also have endospermic seeds). Tobacco (Nicotiana tabaccum), tomato (Solanum lycopersicum), and pepper (Capsicum annuum) are examples of endospermic dicots. In non-endospermic dicots, the triploid endosperm develops normally following double fertilization, but the endosperm food reserves are quickly remobilized and moved into the developing cotyledon for storage. The two halves of a peanut seed (Arachis hypogaea) and the split peas (Pisum sativum) of split pea soup are individual cotyledons loaded with food reserves.

The seed, along with the ovule, is protected by a seed coat that is formed from the integuments of the ovule sac. In dicots, the seed coat is further divided into an outer coat known as the testa and inner coat known as the tegmen.

The embryonic axis consists of three parts: the plumule, the radicle, and the hypocotyl. The portion of the embryo between the cotyledon attachment point and the radicle is known as the hypocotyl (hypocotyl means "below the cotyledons"). The embryonic axis terminates in a radicle (the embryonic root), which is the region from which the root will develop. In dicots, the hypocotyls extend above ground, giving rise to the stem of the plant. In monocots, the hypocotyl does not show above ground because monocots do not exhibit stem elongation. The part of the embryonic axis that projects above the cotyledons is known as the epicotyl. The plumule is composed of the epicotyl, young leaves, and the shoot apical meristem.

Upon germination in dicot seeds, the epicotyl is shaped like a hook with the plumule pointing downwards. This shape is called the plumule hook, and it persists as long as germination proceeds in the dark. Therefore, as the epicotyl pushes through the tough and abrasive soil, the plumule is protected from damage. Upon exposure to light, the hypocotyl hook straightens out, the young foliage leaves face the sun and expand, and the epicotyl continues to elongate. During this time, the radicle is also growing and producing the primary root. As it grows downward to form the tap root, lateral roots branch off to all sides, producing the typical dicot tap root system.

In monocot seeds, the testa and tegmen of the seed coat are fused. As the seed germinates, the primary root emerges, protected by the root-tip covering: the coleorhiza. Next, the primary shoot emerges, protected by the coleoptile: the covering of the shoot tip. Upon exposure to light (i.e. when the plumule has exited the soil and the protective coleoptile is no longer needed), elongation of the coleoptile ceases and the leaves expand and unfold. At the other end of the embryonic axis, the primary root soon dies, while other, adventitious roots (roots that do not arise from the usual place – i.e. the root) emerge from the base of the stem. This gives the monocot a fibrous root system.

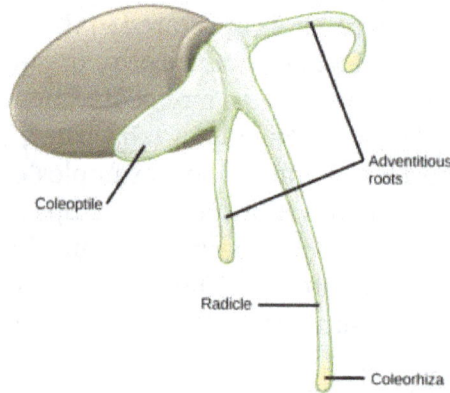

Figure: As this monocot grass seed germinates, the primary root, or radicle, emerges first, followed by the primary shoot, or coleoptile, and the adventitious roots.

Seed Germination

Many mature seeds enter a period of inactivity, or extremely low metabolic activity: a process known as dormancy, which may last for months, years or even centuries. Dormancy helps keep seeds viable during unfavorable conditions. Upon a return to favorable conditions, seed germination takes place. Favorable conditions could be as diverse as moisture, light, cold, fire, or chemical treatments. After heavy rains, many new seedlings emerge. Forest fires also lead to the emergence of new seedlings. Some seeds require vernalization (cold treatment) before they can germinate. This guarantees that seeds produced by plants in temperate climates will not germinate until the spring. Plants growing in hot climates may have seeds that need a heat treatment in order to germinate, to avoid germination in the hot, dry summers. In many seeds, the presence of a thick seed coat retards the ability to germinate. Scarification, which includes mechanical or chemical processes to soften the seed coat, is often employed before germination. Presoaking in hot water, or passing through an acid environment, such as an animal's digestive tract, may also be employed.

Depending on seed size, the time taken for a seedling to emerge may vary. Species with large seeds have enough food reserves to germinate deep below ground, and still extend their epicotyl all the way to the soil surface. Seeds of small-seeded species usually require light as a germination cue. This ensures the seeds only germinate at or near the soil surface (where the light is greatest). If they were to germinate too far underneath the surface, the developing seedling would not have enough food reserves to reach the sunlight.

Development of Fruit and Fruit Types

After fertilization, the ovary of the flower usually develops into the fruit. Fruits are usually associated with having a sweet taste; however, not all fruits are sweet. Botanically, the term "fruit" is used for a ripened ovary. In most cases, flowers in which fertilization has taken place will develop into fruits, and flowers in which fertilization has not taken place will not. Some fruits develop from the ovary and are known as true fruits, whereas others develop from other parts of the female gametophyte and are known as accessory fruits. The fruit encloses the seeds and the developing embryo, thereby providing it with protection. Fruits are of many types, depending on their origin and texture. The sweet tissue of the blackberry, the red flesh of the tomato, the shell of the peanut, and the hull of corn (the tough, thin part that gets stuck in your teeth when you eat popcorn) are all fruits. As the fruit matures, the seeds also mature.

Fruits may be classified as simple, aggregate, multiple, or accessory, depending on their origin. If the fruit develops from a single carpel or fused carpels of a single ovary, it is known as a simple fruit, as seen in nuts and beans. An aggregate fruit is one that develops from more than one carpel, but all are in the same flower: the mature carpels fuse together to form the entire fruit, as seen in the raspberry. Multiple fruit develops from an inflorescence or a cluster of flowers. An example is the pineapple, where the flowers fuse together to form the fruit. Accessory fruits (sometimes called false fruits) are not derived from the ovary, but from another part of the flower, such as the receptacle (strawberry) or the hypanthium (apples and pears).

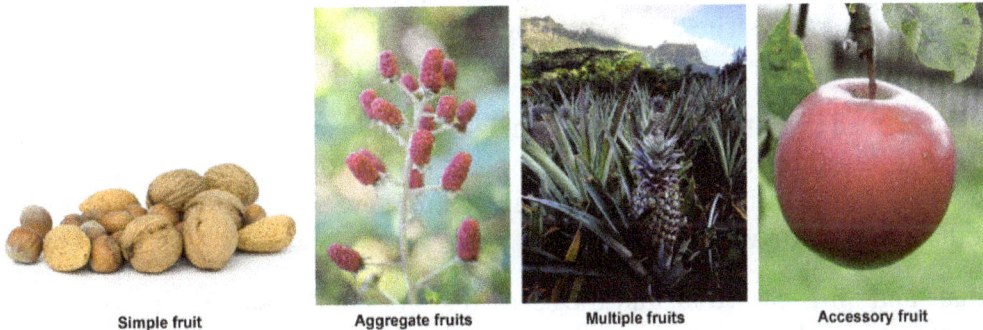

| Simple fruit | Aggregate fruits | Multiple fruits | Accessory fruit |

Figure: There are four main types of fruits. Simple fruits, such as these nuts, are derived from a single ovary. Aggregate fruits, like raspberries, form from many carpels that fuse together. Multiple fruits, such as pineapple, form from a cluster of flowers called an inflorescence. Accessory fruit, like the apple, are formed from a part of the plant other than the ovary.

Fruits generally have three parts: the exocarp (the outermost skin or covering), the mesocarp (middle part of the fruit), and the endocarp (the inner part of the fruit). Together, all three are known as the pericarp. The mesocarp is usually the fleshy, edible part of the fruit; however, in some fruits, such as the almond, the endocarp is the edible part. In many fruits, two or all three of the layers are fused, and are indistinguishable at maturity. Fruits can be dry or fleshy. Furthermore, fruits can be divided into dehiscent or indehiscent types. Dehiscent fruits, such as peas, readily release their seeds, while indehiscent fruits, like peaches, rely on decay to release their seeds.

Fruit and Seed Dispersal

The fruit has a single purpose: seed dispersal. Seeds contained within fruits need to be dispersed far from the mother plant, so they may find favorable and less competitive conditions in which to germinate and grow.

(a)   (b)   (c)

Figure: Fruits and seeds are dispersed by various means. (a) Dandelion seeds are dispersed by wind, the (b) coconut seed is dispersed by water, and the (c) acorn is dispersed by animals that cache and then forget it.

Some fruit have built-in mechanisms so they can disperse by themselves, whereas others require the help of agents like wind, water, and animals. Modifications in seed structure, composition, and size help in dispersal. Wind-dispersed fruit are lightweight and may have wing-like appendages that allow them to be carried by the wind. Some have a parachute-like structure to keep them afloat. Some fruits—for example, the dandelion—have hairy, weightless structures that are suited to dispersal by wind.

Seeds dispersed by water are contained in light and buoyant fruit, giving them the ability to float. Coconuts are well known for their ability to float on water to reach land where they can germinate. Similarly, willow and silver birches produce lightweight fruit that can float on water.

Animals and birds eat fruits, and the seeds that are not digested are excreted in their droppings some distance away. Some animals, like squirrels, bury seed-containing fruits for later use; if the squirrel does not find its stash of fruit, and if conditions are favorable, the seeds germinate. Some fruits, like the cocklebur, have hooks or sticky structures that stick to an animal's coat and are then transported to another place. Humans also play a big role in dispersing seeds when they carry fruits to new places and throw away the inedible part that contains the seeds.

All of the above mechanisms allow for seeds to be dispersed through space, much like an animal's offspring can move to a new location. Seed dormancy, which was described earlier, allows plants to disperse their progeny through time: something animals cannot do. Dormant seeds can wait months, years, or even decades for the proper conditions for germination and propagation of the species.

Isogamy

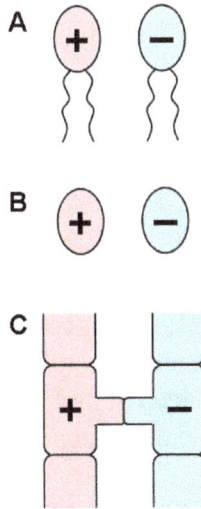

The simplest type is isogamous sexual process (isogamy). In this case, both gametes are mobile and are absolutely identical in size and appearance. During the fusion they retain mobility by flagella at the front end of rounded or pear-shaped body. Isogamy occurs in lower plants — algae, fungi.

Heterogamy

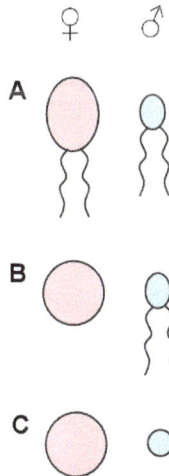

Heterogamy is also known as Anisogamy. In anisogamy both gametes are mobile, they have flagella, but differ in size. One of them, the smaller, has more mobility and is considered to be a male gamete. Another one, a little larger, is less mobile. In its plasma, there is a certain amount of reserve nutrients. It is considered to be a female gamete.

Anisogamy occurs only in lower plants, such as in some green algae (Chlamydomonas Braunii) and brown algae (e.g. Ectocarpus).

The gametes are usually produced in special cells that sometimes do no differ from conventional plants' vegetative cells, but more often have different distinctive form in comparison with the latter. The cells that produce gametes are called gametangia.

Oogamy

In oogamous sexual process (oogamy), gametes sharply differ in shape, size and character. One of them — male — is very shallow, its protoplast consists mainly of the nucleus and a thin layer of the cytoplasm; in front of it there is a small amount of plasma, from which depart flagella. Male gamete is actively mobile and has a special name — spermatozoid. Female gamete is deprived of motility. It's called an egg or ovum. Its major protoplast has a large nucleus. The cytoplasm contains a large amount of nutrients.

Oogamy occurs in all higher plants and in vast majority of lower plants. Male organs, in which develop spermatozoids, are called antheridia; Female, in which form the egg cells, in lowest plants are called oogonia, in higher plants — archegonium. Oogonium is always a unicellular form; archegonium is already a multicellular body having a complex structure.

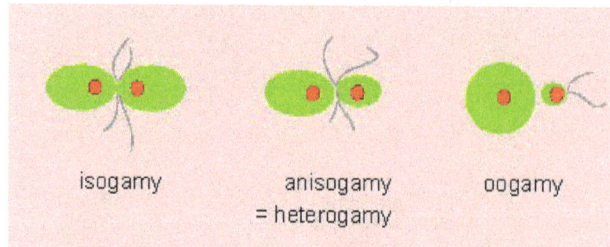

isogamy   anisogamy = heterogamy   oogamy

Amphimixis

In plants amphimixis concept can be considered identical to the concept of cross-pollination. This method of fertilization shows the main advantage of sexual reproduction: each individual becomes a genetically unique, as it contains a specific combination of genes.

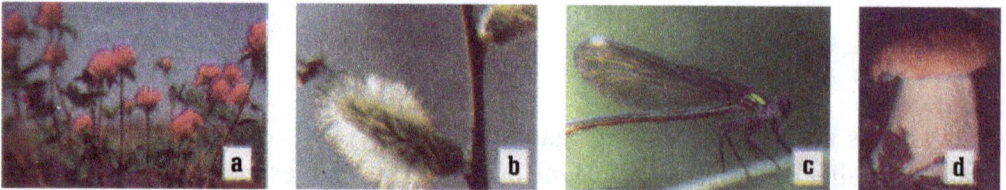

Figure: Some common species of flora and fauna, reproducing via amphimixis:
a — red clover; b — goat willow; c — dragonfly banded demoiselle; d — cep (boletus)Et re nis expediam ipsuntibust, qui omnis maximilis doloreperio event, et aut dolorectur mil ipsame cust, sed maximus nonsendi as am dolora quibus natium harchil il iducia dus.

Obit, cullupi eturit et endam, eosam sam rerenimus, ut et moluptatur, cum incienis inulpa arum escilition reperunte se volupta temporest rehenditio ipsunt re et omnis dolorerum evenihilique eremquam que volumqui culliate dolorum eaquis dolendi inihitatius eatur?

Fugit, cus ulparum illique rem aborestem volorrumque endebit atectisci di volorernati corest que nonsequia eumet voluptatur, none suntiuntium quas eostios inullup tatium consequo tota est millupt assus, in porisquidi dollant laut pore reperru ptatque noneces dolestrupta cusant quat est, offictur?

Ossi dolorae si rent volest, cum ipicia que pliqui consedi dolecepe dolorem latem ligent utemos dest.

References

- Bicknell; Anna M. Koltunow (2004). "Understanding Apomixis: Recent Advances and Remaining Conundrums". The Plant Cell. 16 (suppl 1): S228–S245. doi:10.1105/tpc.017921. PMC 2643386

- Asexual-reproduction, boundless-biology: lumenlearning.com, Retrieved 15 July 2018

- Savidan, Y.H. (2000). "Apomixis: genetics and breeding". Plant Breeding Reviews. 18: 13–86. doi:10.1002/9780470650158.ch2

- Parthenocarpy-information, fegen, fruits: gardeningknowhow.com, Retrieved 21 July 2018

- Sexual-reproduction-in-plants-(isogamy-anisogamy-oogamy): worldofschool.org, Retrieved 20 May 2018

- Steil, W.N. (1939). "Apogamy, apospory, and parthenogenesis in the Pteridophytes". The Botanical Review. 5 (8): 433–453. doi:10.1007/bf02878704

- Sexual-reproduction, sexual-reproductio-in-flowering-plants: toppr.com, Retrieved 25 June 2018

- Pichot, C.; El Maataoui, M.; Raddi, S.; Raddi, P. (2001). "Conservation: Surrogate mother for endangered Cupressus". Nature. 412 (6842): 39. doi:10.1038/35083687

- Sexual-reproduction-in-plants, suny-wmopen-biology: lumenlearning.com, Retrieved 29 March 2018

Chapter 4

Crop Improvement: Genetic Concepts and Principles

In agriculture, crops are modified using genetic engineering. A new trait is introduced or some existing trait is improved for better agricultural productivity. The topics elaborated in this chapter such as mitosis, meiosis, structural chromosomal, chromosome number, chromosomal crossover, gene interaction, cytoplasmic inheritance, pleiotropy, mutations, etc. will help in providing a better understanding of the role of genetics for enhanced crop production.

Mitosis and Meiosis

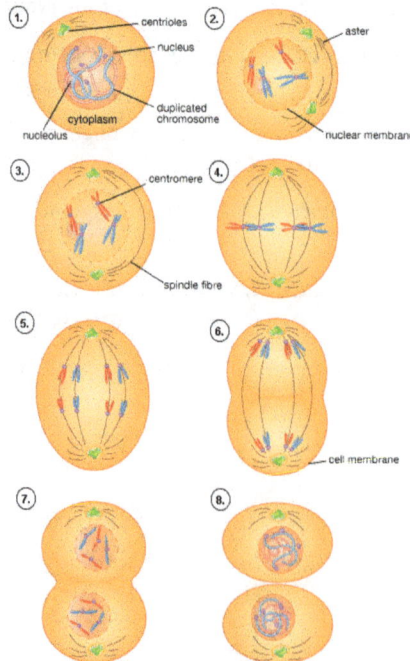

Mitosis

Organisms must be able to grow and reproduce. Prokaryotes, such as bacteria, duplicate deoxyribonucleic acid (DNA) and divide by splitting in two, a process called binary fission. Cells of eukaryotes, including those of animals, plants, fungi, and protists, divide by one of two methods: mitosis or meiosis.

Mitosis produces two cells, called daughter cells, with the same number of chromosomes as the parent cell, and is used to produce new somatic (body) cells in multicellular eukaryotes or new individuals in single-celled eukaryotes. In sexually reproducing organisms, cells that produce gametes (eggs or sperm) divide by meiosis, producing four cells, each with half the number of chromosomes possessed by the parent cell.

Chromosome Replication

All eukaryotic organisms are composed of cells containing chromosomes in the nucleus. Chromosomes are made of DNA and proteins. Most cells have two complete sets of chromosomes, which occur in pairs. The two chromosomes that make up a pair are homologous, and contain all the same loci (genes controlling the production of a specific type of product).

These chromosome pairs are usually referred to as homologous pairs. An individual chromosome from a homologous pair is sometimes called a homolog. For example, typical lily cells contain twelve pairs of homologous chromosomes, for a total of twenty-four chromosomes.

Cells that have two homologous chromosomes of each type are called diploid. Some cells, such as eggs and sperm, contain half the normal number of chromosomes (only one of each homolog) and are called haploid. Lily egg and sperm cells each contain twelve chromosomes.

DNA must replicate before mitosis or meiosis can occur. If daughter cells are to receive a full set of genetic information, a duplicate copy of DNA must be available. Before DNA replication occurs, each chromosome consists of a single long strand of DNA called a chromatid. After DNA replication, each chromosome consists of two chromatids, called sister chromatids.

The original chromatid acts as a template for making the second chromatid; the two are therefore identical. Sister chromatids are attached at a special region of the chromosome called the centro mere. When mitosis or meiosis starts, each chromosome in the cell consists of two sister chromatids.

Mitosis andmeiosis produce daughter cells with different characteristics. When a diploid cell undergoes mitosis, two identical diploid daughter cells are produced.

When a diploid cell undergoes meiosis, four unique haploid daughter cells are produced. It is important for gametes to be haploid, so that when an egg and sperm fuse, the diploid condition of the mature organism is restored.

Cellular Life Cycles

Mitosis and meiosis occur in the nuclear region of the cell, where all the cell's chromosomes are found. Nuclear control mechanisms begin cell division at the appropriate time.

Some cells rarely divide by mitosis in adult organisms, while other cells divide constantly, replacing old cells with new. Meiosis occurs in the nuclei of cells that produce gametes. These specialized cells occur in reproductive organs, such as flower parts in higher plants.

Cells, like organisms, are governed by life cycles. The life cycle of a cell is called the cell cycle. Cells spend most of their time in interphase. Interphase is divided into three stages: first gap (G1), synthesis (S), and second gap (G2).

During G1, the cell performs its normal functions and often grows in size. During the S stage, DNA replicates in preparation for cell division. During theG2 stage, the cell makes materials needed to produce the mitotic apparatus and for division of the cytoplasmic components of the cell.

At the end of interphase, the cell is ready to divide. Although each chromosome now consists of two sister chromatids, this is not apparent when viewed through a microscope. This is because all the chromosomes are in a highly relaxed state and simply appear as a diffuse material called chromatin.

Mitosis

Mitosis consists of five stages: prophase, prometaphase, metaphase, anaphase, and telophase. Although certain events identify each stage, mitosis is a continuous process, and each stage gradually passes into the next. Identification of the precise state is therefore difficult at times.

During prophase, the chromatin becomes more tightly coiled and condenses into chromosomes that are clearly visible under a microscope, the nucleolus disappears, and the spindle apparatus begins to form in the cytoplasm.

In prometaphase, the nuclear envelope breaks down, and the spindle apparatus is now able to invade the nuclear region. Some of the spindle fibers attach them selves to a region near the centromere of each chromosome called the kinetochore.

The spindle apparatus is the most obvious structure of the mitotic apparatus. The nuclear region of the cell has opposite poles, like the North and South Poles of the earth. Spindle fibers reach from pole to pole, penetrating the entire nuclear region.

During metaphase, the cell's chromosomes align in a region called the metaphase plate, with the sister chromatids oriented toward opposite poles. The metaphase plate traverses the cell, much like the equator passes through the center of the earth. Sister chromatids separate during anaphase.

The sister chromatids of each chromosome split apart, and the spindle fibers pull each sister chromatid (now a separate chromosome) from each pair toward opposite poles, much as a rope-tow pulls a skier up a mountain. Telophase begins as sister chromatids reach opposite poles. Once the chromatids have reached opposite poles, the spindle apparatus falls apart, and the nuclear membrane re-forms.

Meiosis

Meiosis: Selected Phases

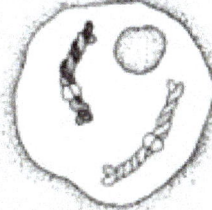

(1) Early prophase II   (2) Prophase I

(3) Late prophase I   (4) Metaphase I

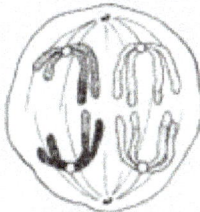
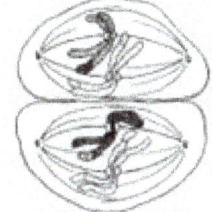

(5) Amaphase I   (6) Metaphase III

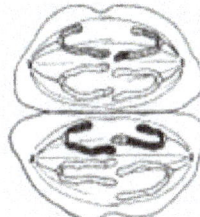
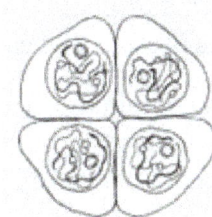

(7) Anaphase II   (8) Late telophase III

Meiosis is a more complex process than mitosis and is divided into two major stages: meiosis I and meiosis II. As in mitosis, interphase precedes meiosis. Meiosis I consists of prophase I, metaphase I, anaphase I, and telophase I.

Meiosis II consists of prophase II, metaphase II, anaphase II, and telophase II. In some cells, an interphase II occurs between meiosis I and meiosis II, but no DNA replication occurs.

During prophase I, the chromosomes condense, the nuclear envelop falls apart, and the spindle apparatus begins to form. Homologous chromosomes come together to form tetrads (a tetrad consists of four chromatids, two sister chromatids for each chromosome). The arms of the sister chromatids of one homolog touch the arms of sister chromatids of the other homolog, the contact points being called chiasmata.

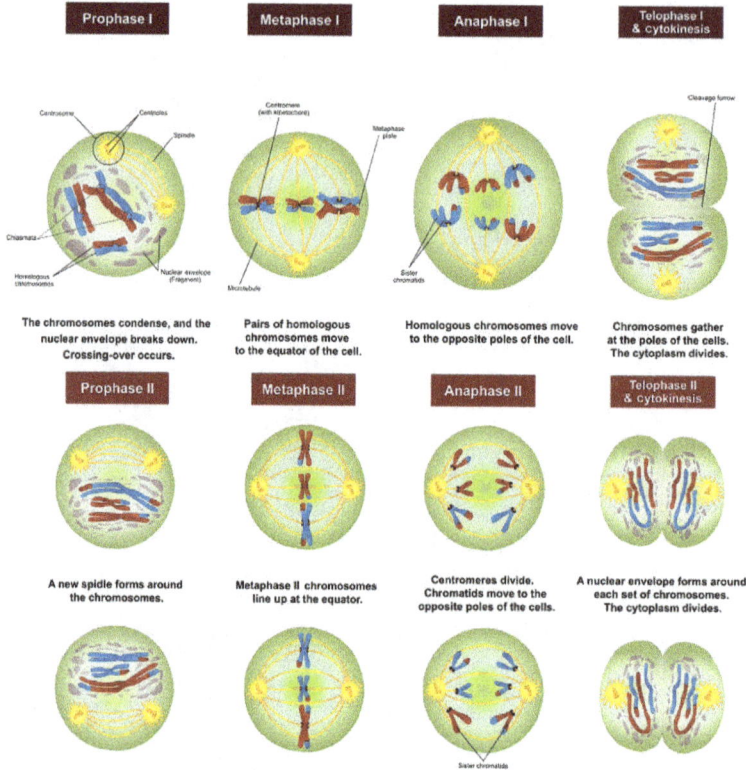

Prophase I	Metaphase I	Anaphase I	Telophase I & cytokinesis
The chromosomes condense, and the nuclear envelope breaks down. Crossing-over occurs.	Pairs of homologous chromosomes move to the equator of the cell.	Homologous chromosomes move to the opposite poles of the cell.	Chromosomes gather at the poles of the cells. The cytoplasm divides.
Prophase II	Metaphase II	Anaphase II	Telophase II & cytokinesis
A new spidle forms around the chromosomes.	Metaphase II chromosomes line up at the equator.	Centromeres divide. Chromatids move to the opposite poles of the cells.	A nuclear envelope forms around each set of chromosomes. The cytoplasm divides.

Each chiasma represents a place where the arms have the same loci, so called homologous regions. During this intimate contact, the chromosomes undergo crossover, in which the chromosomes break at the chiasmata and swap homologous pieces.

This process results in recombination (the shuffling of linked alleles, the different forms of genes, into new combinations), which results in increased variability in the off spring and the appearance of character combinations not present in either parent.

Tetrads align on the metaphase plate during metaphase I, and one spindle fiber attaches to the kinetochore of each chromosome. In anaphase I, instead of the sister chromatids separating, they remain attached at their centromeres, and the homologous chromosomes separate, each homolog from a tetrad moving toward opposite poles.

Telophase I begins as the homologs reach opposite poles, and similar to telophase of mitosis, the spindle apparatus falls apart, and a nuclear envelope re-forms around each of the two haploid nuclei. Because the number of chromosomes in each of the telophase

I nucleus is half the number in the parent nucleus, meiosis I is sometimes called the reduction division.

Meiosis II is essentially the same as mitosis, dividing the two haploid nuclei formed in meiosis I. Prophase II, metaphase II, anaphase II, and telophase II are essentially identical to the stages of mitosis. Meiosis II begins with two haploid cells and ends with four haploid daughter cells.

Nuclear Division and Cytokinesis

Mitosis and meiosis result in the division of the nucleus. Nuclear division is nearly always coordinated with division of the cytoplasm. Cleaving of the cytoplasmto form new cells is called cytokinesis. Cytokinesis begins toward the middle or end of nuclear division and involves not just the division of the cytoplasm but also the organelles.

In plants, after nuclear division ends, a new cell wall must be formed between the daughter nuclei. The new cell wall begins when vesicles filled with cell wall material congregate where the metaphase plate was located, producing a structure called the cell plate.

When the cell plate is fully formed, cytokinesis is complete. Following cytokinesis, the cell returns to interphase. Mitotic daughter cells enlarge, reproduce organelles, and resume regular activities. Following meiosis, gametes may be modified or transported in the reproductive system.

Alternation of Generations

Meiotic daughter cells continue development only if they fuse during fertilization. Mitosis and meiosis alternate during the life cycles of sexually reproducing organisms. The life-cycle stage following mitosis is diploid, and the stage following meiosis is haploid. This process is called alternation of generations.

In plants, the diploid state is referred to as the sporophyte generation, and the haploid stage as the gametophyte generation. In nonvascular plants, the gametophyte generation dominates the life cycle. In other words, the plants normally seen on the forest floor are made of haploid cells.

The sporophytes, which have diploid cells, are small and attached to the body of the gametophyte. In vascular plants, sporophytes are the large, multicellular individuals (such as trees and ferns), whereas gametophytes are very small and either are embedded in the sporophyte or are free-living, as are ferns.

The genetic variation introduced by sexual reproduction has a significant impact on the ability of species to survive and adapt to the environment. Alternation of generations allows sexual reproduction to occur, without changing the chromosome number characterizing the species.

Plant Chromosome

Chromosomes contain the genetic information of cells. Replication of chromosomes assures that genetic information is correctly maintained as cells divide.

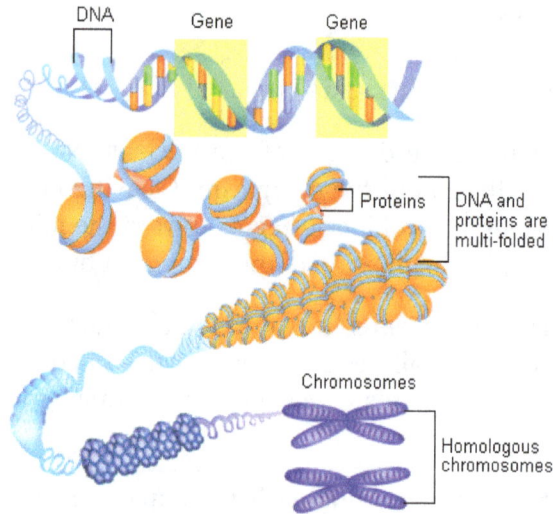

The genome of an organism is the sum total of all the genetic information of that organism. In eukaryotic cells, this information is contained in the cell's nucleus and organelles, such as mitochondria and plastids. In prokaryotic organisms (bacteria and archaea), which have no nucleus, the genomic information resides in a region of the cell called the nucleoid.

A chromosome is a discrete unit of the genome that carries many genes, or sets of instructions for inherited traits. Genes, the blueprints of cells, are specific sequences of deoxyribonucleic acid (DNA) that code for messenger ribonucleic acids (monas), which in turn direct the synthesis of proteins.

Each eukaryotic chromosome contains a single long DNA molecule that is coiled, folded, and compacted by its interaction with chromosomal proteins called histone. This complex of DNA with chromosomal proteins and chromosomal RNAs is chromatin.

DNA of higher eukaryotes is organized into loops of chromatin by attachment to a nuclear scaffold. The loops function in the structural organization of DNA and may increase transcription of certain genes by making the chromatin more accessible.

To maintain the genetic information of a cell, it is essential that chromosomes correctly replicate and divide as a cell divides. After DNA replication, chromosomes separate in a process called mitosis.

During this process, the nuclear envelope breaks down and chromosomes condense into compact structures. A cellular structure known as the mitotic spindle forms, pulling pairs of replicated chromosomes apart so that the two cells receive identical sets of chromosomes.

Chromosomes are readily visualized when they condense during cell division. All the chromosomes of a cell visualized during mitosis constitute that cell's karyotype.

Each chromosome has a centromere—a constricted area of the condensed chromosome where the mitotic or meiotic spindle attaches to assure correct distribution of chromosomes during cell division—and a telomere, the end or tip of a chromosome, which contains tandem repeats of a short DNA sequence.

The number of chromosomes in a gamete (either egg or sperm) is the haploid number, n. The haploid number of chromosomes in humans is 23; in corn, 10; in peas, 7; in Arabidopsis (the model organism used in much botanical research), 4.

Some carp and some ferns have more than 50 chromosomes in the haploid genome. Pollen grains of some plants, such as pear, contain three haploid cells: One directs the growth of the pollen tube down the style to the ovary; the other two are sperm.

In flowering plants (angiosperms), there is a unique double fertilization where by one sperm nucleus fuses with the egg nucleus to form the diploid (2n) zygote, and the other sperm nucleus fuses with two polar nuclei to form the triploid nutritive tissue, or endosperm, which will nourish the embryo in the seed.

The zygote then increases in cell number by mitosis, a type of cell division during which chromosomes in a nucleus are replicated and then separated to form two genetically identical daughter nuclei.

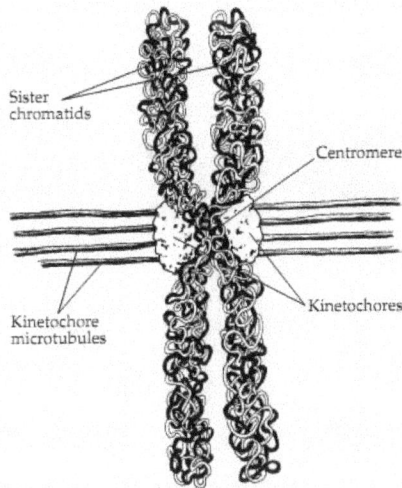

Schematic of a chromosome

This is followed by cytokinesis, the process of cytoplasmic division, which results in two daughter cells, each having the same number of chromosomes and genetic composition as the parent cell. The mature 2n plant forms the haploid (n) gametes by meiosis, a type of cell division that reduces the number of chromosomes to the haploid number.

A distinctive feature of plant cell division is the plant cell has three genomes (the nuclear, mitochondrial, and plastid genome) to replicate and divide. The chromosomes of eukaryotes consist of unique genes among a complex pattern of repeated DNA sequences. Arabidopsis has only 4 chromosomes containing about 120 million base pairs.

There are typically between twenty and one hundred copies of the mitochondrial genome per mitochondrion, ranging in size from two hundred to twenty-four hundred kilo base pairs (or kb; one kilo base pair equals one thousand base pairs).

Plant mitochondrial genomes are much larger than the mitochondrial genomes of yeast. Chloroplast genomes range in size from 130 to 150 kb, with 50 to 150 copies of that genome per plastid.

In cell division in plant cells, the two daughter nuclei are partitioned to form two separate cells by a cell plate that grows at the equator of the mother cell.

DNA synthesis occurs in the synthesis (S) phase, beginning at origins of replication distributed around the genome, occurring on average every 66 kb in dicotyledonous plants and on average every 47 kb in monocotyledonous plants.

Heterochromatin is the term for regions of chromosomes that are permanently in a highly condensed state, are not transcribed, and are late-replicating.

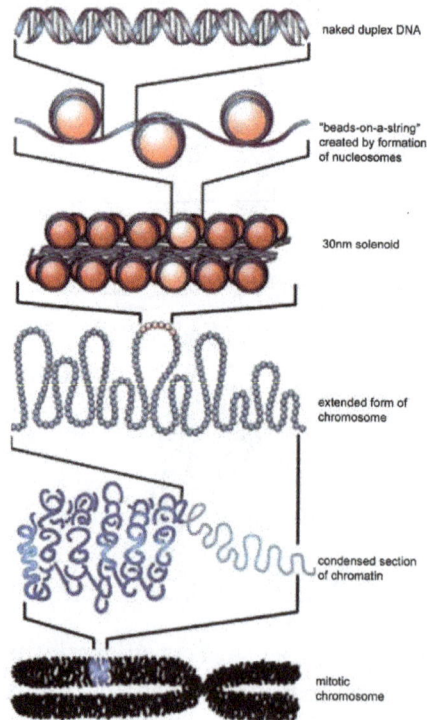

Heterochromatin contains highly repeated DNA sequences. Euchromatin is the rest of the chromosomes that is extended, accessible to RNA polymerase, and at least partially transcribed.

Some plants have extra chromosomes that do not seem to be essential. These are called accessory or supernumerary chromosomes. They have been most studied in corn where these extra chromosomes are called B-chromosomes. B-chromosomes are usually highly condensed heterochromatin that may or may not be present in an individual of that species.

An increase in the copy number of the genome is common in plants, occurring during the development of individuals. Polyploids have three or more complete sets of chromosomes in their nuclei instead of the two sets found in diploids. For example, in Arabidopsis, tissues of increasing age have an increase in polyploidy, reaching up to sixteen duplications.

Chromosome Number

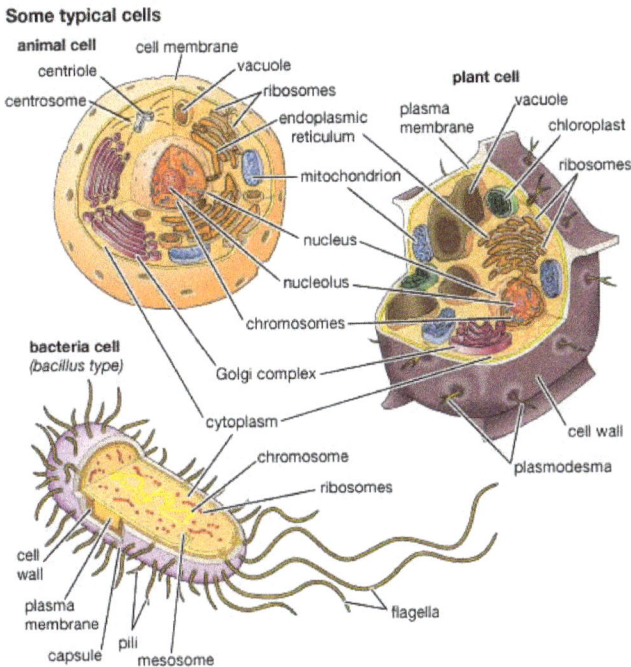

Normally the number of the chromosomes is constant for all individuals of a particular species and this is of great importance in determining the phylogeny and taxonomy of the species. Sometimes closely related species have similar chromosomes number. The number and the set of chromosomes of the gametic cells, such as sperms and ova is known as the gametic, reduced or haploid (n) and the set. The haploid set of chromosome is known as genome. Genome is the sum total of the genes in the haploid set of chromosomes. The somatic or body cells of most of the of the organisms contain two haploid sets genomes received from the parent are known as the diploid (2n) cells. The diploid cells achieve the diploid condition of the chromosomes by the union of the haploid male and female gametes by way of sexual reproduction. In many species the somatic cells usually contains more than two sets of chromosomes. Such cells (as well as plants) are termed triploids (3n) having three sets, tetraploid (4n) having four sets etc. The condition in which chromosome are present in exact multiples of 'n' sets is called euploidy. When change in chromosome number does not involve exact multiple of 'n' set but only a few of the chromosomes, the condition is termed aneuploidy.

Example: monosomic (2n-1), trisomic (2n+1), nullisomic (2n-2), tetrasomic (2n+2). The euploidy and aneuploidy together constitute polyploidy, a condition in which chromosome number is increased. The haploid chromosome number recorded in plants is only two (n=2). The highest chromosome number recorded in plant is in Ophioglossum reticulatum (Pteridophyta) where the diploid number (2n) is as much as 1260.

Chromosome Number (2n) of some Common Plants

1.	Aspergillus = 8 or 16	2.	Neurospora = 14
3.	Penicillum = 4	4.	Yeast = 4 or 8
5.	Chlamydomonas = 16	6.	Watermelon =22
7.	Cucumber = 14	8.	Papaya =18
9.	Pear =34, 51, 68	10.	Peanut = 40
11	Coffee = 44	12.	Sunflower =34
13.	Banana = 22, 44, 55, 77, 88	14.	Bean =22
15.	Orange = 18 27 36	16.	Sugarcane = 80
17.	Squash = 40	18.	Pea =14
19.	Onion= 16	20.	Cabbage = 18
21.	Radish = 18	22.	Tomato = 24
23.	Tobacco = 48	24.	Potato =48
25.	Paddy =24	26.	Tea =30
27.	Wheat = 42	28.	Maize = 20
29.	Apple =34	30.	Cotton =52
31.	Barley = 14	32.	Avena = 42

Linkage

In every plants, there are present thousands of genes. The number of genes is more as compared to the number of chromosomes e.g., pea. It is clear that every chromosome possesses more than one gene. In 1903, William Sutton had expressed the possibility that every chromosome must have more than one unit factors.

Linkage may be complete, incomplete, or absent (not detectable), depending upon the distance between linked genes in a chromosome.

W. Bateson, Saunders and Punnett during experiments, found that the result obtained from a cross in sweet pea show a deviation from law of independent assortment

They were crossed the plant Lathyrus odoratus (sweet pea). In this plant, purple color of flower is dominant over red color and long pollen are dominant over round pollen. The purple flowers (B) and long pollen (L) were crossed with red flowers (b) and round pollen (l), in F1 generation the plants (BbLl) produced purple flower and long pollen, as expected. These plants were crossed with plants having red flowers and round pollen (bbll)

The resulting genotypes, and their actual and expected numbers under independent assortment, were as follows:

Phenotype	Genotype	Observed	Expected
Purple, long	B_L_	284	215
Purple, round	B_ll	21	71
Red, long	bbL_	21	72
Red, round	bbll	55	24
Total		381	381

In F2 generation 1 : 1 : 1 : 1 ratio was expected after testcross but 7 : 1 : 1 : 7 ratio was actually obtained. This indicates that there is a tendency in dominant alleles to remain together. And similar is the case with recessive alleles. This deviation was, termed as coupling by Bateson. It was also observed that when two such dominant alleles or two recessive alleles come from different parents, they tend to remain separate and was termed as repulsion. When the plants with purple flowers and round pollen (Bbll) were crossed with red flowers and long pollen (bbll). The results of testcross were similar to that coupling phase 1 : 7 : 7 : 1 ratio instead of expected 1 : 1 : 1 : 1. Therefore Bateson and co-workers coined two words as coupling and repulsion. Bateson and Punnett could not explain the exact reasons of coupling and repulsion, and it was T.H. Morgan who while performing experiments with Drosophila, in 1910, found that coupling or repulsion was not complete.

Linked genes: By studying the inheritance of characters in the fruitfly Drosophila, TH Morgan and colleagues (1910) determined that genes are not completely independent as Mendel had thought, but that they tend to be inherited in groups. Since independent assortment does not occur, a dihybrid cross following two linked genes will not produce and F2 phenotypic ratio of 9:3:3:1. They observed that genes in the same chromosome are often transmitted together as a group, but that this was not always so and that 'crossing-over' between chromosomes could occur to disrupt these linkage groups. Genes that are present on the same chromosome, and that tend to be inherited (transmitted to the gametes) together, are termed linked genes because the DNA sequence containing the genes is passed along as a unit during meiosis. The closer that genes reside on a particular chromosome, the higher the probability that they will be inherited as a unit, since crossing over between two linked genes is not as common. The genes present on same chromosome, thus, would not assort (separated) independently. Such type of genes are called linked genes and this phenomenon is called linkage.

For example, the "A" and "B" alleles which are present in same chromosome will both be passed on together if the chromosome is inherited. "A" and "B" are linked due to their occurrence in the same chromosomes. Similarly, "a" and "b" are linked in the other chromosome.

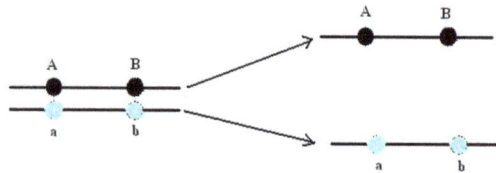

Linked genes tend to be inherited together because they are located on the same chromosome

Morgan defined linkage as follows: "that the pairs of genes of homozygous parents tend to enter in the same gametes and to remain together, whereas same genes from heterozygous parent tend to enter in the different gametes and remain apart from each other. He further stated that the tendency of linked genes remaining together in original combination is due to their location in the same chromosome." According to him the degree or strength of linkage depends upon the distance between the linked genes in the chromosome.

Chromosomes Theory of Linkage:- Morgan along with castle formulated the chromosome theory of linkage which is as follows :-

1. The genes which show the phenomenon of linkage are situated in the same chromosomes and separated during the process of inheritance.

2. The distance between the linked genes determines the strength of linkage. The closely located genes show strong linkage than the widely located genes which show the weak linkage.

3. The genes are arranged in linear fashion in the chromosomes.

Types of linkage:- Linkage may be complete, incomplete, or absent (not detectable) linkage, depending upon the distance between linked genes in a chromosome.

Complete linkage

During synapsis, exchange of segments takes place. In such condition the possibility of separation of two genes situated close together is greatly reduced. When genes are closely associated and tend to transmit together, it is called complete linkage.

A   B

"A" and "B" are too close to each other.

Incomplete linkage

When linked genes are situated at long distance in chromosomes and have chances of separation by crossing over are called incompletely linked genes and phenomenon of their inheritance is called incomplete linkage.

A   B

"A" and "B" are separated to allow crossing-over between them

Absent (not detectable) linkage

The probability of crossover increases with the physical distance between genes on a chromosome, and genes that are located quite far From each other within a linkage group may not exhibit any detectable linkage.

A   B

"A" and "B" are too far apart to allow crossing-over between them in all

Linkage groups:- The group of linked genes that are located on the same chromosome, called linkage groups. Because, all the genes of a chromosome have their identical genes (alleomorphs) on the homologous chromosome, therefore linkage groups of a homologous pair of chromosome is considered as one. In any species, the number of linkage groups is equal to the number of pairs of chromosomes. e.g. Corn (Zea mays) has 10 pairs of chromosomes and 10 linkage groups.

Arrangement of Genes

Cis- arrangement of genes: If the dominant alleles (A,B) of two linked genes are present on the same chromosomes and their recessive alleles (a, b) are present on the homologous chromosomes the arrangement of genes is called cis-arrangement.

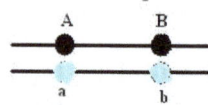

Cis- arrangement of genes

Trans-arrangement if one dominant gene and other recessive gene present on one chromosome (A, b) and their allele type (a, B) on the chromosome this type of arrangement is known as trans-arrangement.

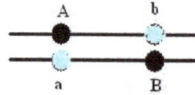

Trans-arrangement of genes

Example of Linkage

Linkage in Maize

The phenomenon of linkage can be easily demonstrated in maize, as recombination also can be seen and traits in the seeds can be easily observed. Each ear having hundreds of seeds makes the observation easy.

Hutchison has demonstrated linkage in two varieties of maize. One had colored and full seeds, while the other had colorless shrunken seeds. The gene for color is C and/is dominant over the colorless condition c.

Similarly the full seed F is dominant over shrunken condition. The genotypes of the parents which represent the pure line are CCFF and ccff. In a cross between the two, as expected the F is colored and full. The genotype is CcFf.

The genotype however is written as CF/cf to indicate C & F and c &f are linked. The gene combinations are written in the manner in which they enter the zygote. If the genotype is- written as CcFf, it is indicative of independent assortment.

If the F, follows the Mendelian pattern of segregation, four kinds of gametes should be -produced in equal proportions. These are CF, Cf cF, and cf. In order to find out, whether these gametes are formed in equal numbers a test cross in done and the F dihybrid is crossed with the double recessive parent (CF/cfx cf/cf).

If the F, dihybrid forms the 4 types of gametes in equal proportions it should yield the four classes of progenies. (Colored full, colored shrunken, colorless full and colorless shrunken) in the ratio of 1:1:1:1. But the actual numbers of progenies obtained are as given in the following table.

The parental combinations namely colored full and colorless shrunken are in large numbers among the progeny than the recombination viz., colored shrunken and colorless full.

If the assortment were to be independent all the four categories should have been produced in equal proportions. Instead, parental combinations abound. Out of a total of 8368 individuals about 96.4% (4032 + 4035) are parental combinations and only about 3.6% (149 + 152) are recombination.

This shows clearly genes C and F have not assorted independently, so also c and f. In other words, the linkage of 96.4%, while recombination i.e., break of linkage is seen in only 3.4% of the progeny.

In another cross in maize plant dealing with the same characters, different parental combinations were selected. The two parents selected for crossing were colored shrunken and colorless full.

Even in this cross, as the table below shows parental combinations were more than the recombination. But here the parental combinations are different from those of the first experiment.

This experiment shows clearly, that whatever parental characters are there, they tend to be inherited together and not assorted independently.

This example in maize, however is not an example of complete linkage, because at least in some instances recombination is seen indicating the break in linkage.

While it is true linkage is there in 96.4% and 97.06% cases in the above two experiments, it is also true that the genes have assorted independently in at least 3.6% and 2.94% cases.

Thus the type of linkage seen in maize is called incomplete linkage. In a complete linkage, parental combinations are retained in 100%, there is just no recombination.

Chromosomal Crossover

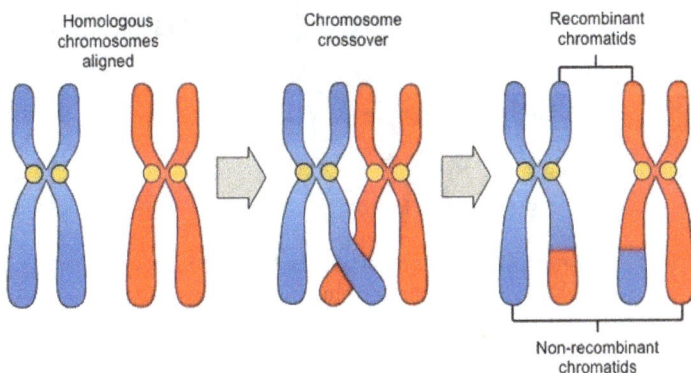

Homologous chromosomes aligned — Chromosome crossover — Recombinant chromatids — Non-recombinant chromatids

Crossing-over takes place during prophase I of meiosis. Synapsis is the process in which homologous chromosomes carefully pair. The pairing allows for an orderly first division to send one chromosome from each pair to separate cells. The close association of the homologous chromosomes also allows for crossing over between non-sister chromatids. During this process sections of the chromosomes break off and are exchanged between non-sister chromatids. When non-sister chromatids crossover, chromatids can be made that have a new combination of genes compared to the original combination on the chromosome. The original combination was inherited from the organism's parents and is called the parental combination of genes. The new combination made is called the recombinant combination. In figure, a crossover occurs but the original or parental combination of CS (red and plump) and cs (white and shrunken) will stay together. Crossing over can cause new gene combinations to occur on a chromosome if the crossover occurs between the linked genes.

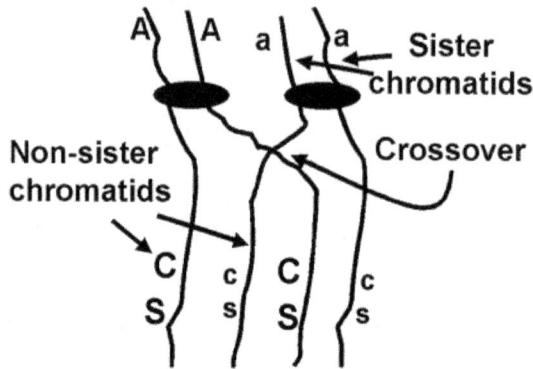

Figure: Crossover not between C,c and S,s.
Only CS and cs. Parental gametes are made.

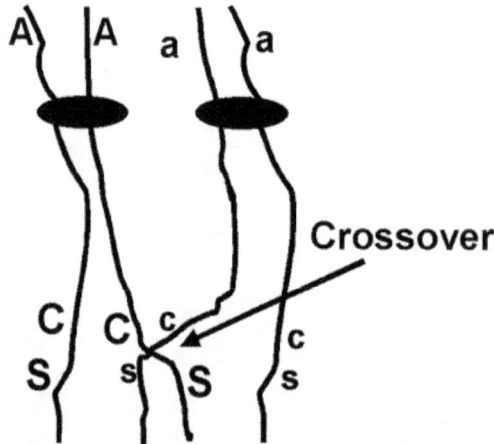

Figure: Crossover between the C,c and S,s gives
Cs and cS recombinant plus CS and cs parental.

When a crossover occurs between genes, chromatids with both the parental combination and chromatids with a new combination will be made. We can see this in figure.

Two of the chromatids are not involved in the crossing over. These chromatids will maintain the parental combination and when meiosis is complete, the two gametes made that have these chromosomes will be called parental gametes. The gametes made that have the other two chromosomes, those that went through crossing over and have the new gene combination, are called recombinant gametes.

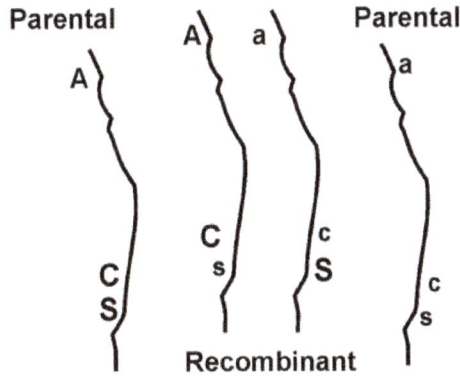

Figure: Parental and recombinant gametes from crossovers between C,c and S,s.

When crossing over occurs between two non-sister chromatids, cells will make equal numbers of recombinant and parental gametes.

Types of Crossing Over

Crossing overs are of many types depending on number of chiasma:-

1. Single crossing over:- When the chiasma formation takes place at a single point of The chromosome pair this type of crossing over is known as single crossing over. In this types two crossed over chromatids and two non crossed over chromatids are formed.

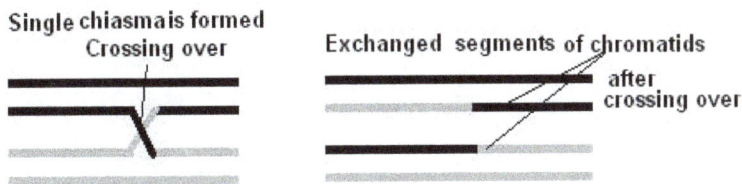

Figure: Single crossing over

2. Double crossing over:- When the chiasmata occur at two places in the same chromosomes known as double crossing over In the double crossing over formation of each chiasma is independent of the other and in it four types of recombination is possible.

Two types of chiasma may be formed in double cross over:-

   o   Reciprocal chiasma in this type both the chiasma are formed on two same chromatids. So, the second chiasma restores the order which was changed by the first Chiasma, and as a result two non- cross over chromatids are formed.

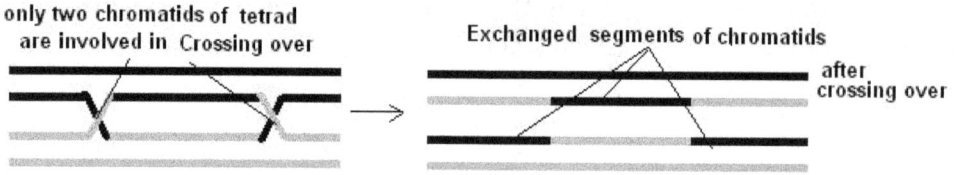

Figure: Double crossing over (Reciprocal chiasma)

In this type out of four chromatids only two are involved in the double crossing over.

- o Complimentary chiasma When both the chromatids taking part in the second chiasma are different from those chromatids involved in the first. In this type four single cross overs are produced but no non cross over. Complimentary chiasma occurs when three or four chromatids of tetrad undergo crossing over.

3. Multiple crossing over When crossing over take place at more than two point in the same chromosome pair it is known as multiple crossing over. It occurs rarely.

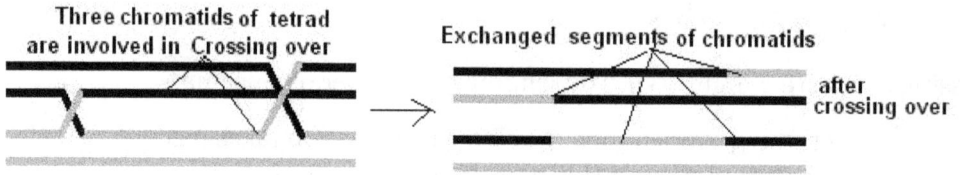

Figure: Double crossing over (Complimentary chiasma)

There are two theories on the physical nature of the process:-

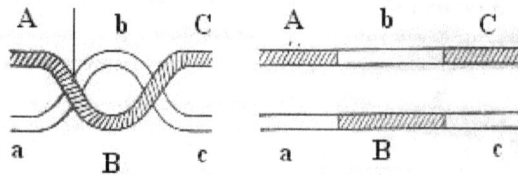

Showing chiasma formation and crossing over - Classical theory

1. Classical theory or two plane theory (L. W. Sharp):- proposes that cross-over and formation of the chiasma occur first, followed by breakage and reunion with the reciprocal homologues. According to this theory, chiasma formation need not be accompanied by chromosome breakage. But this theory was not accepted.

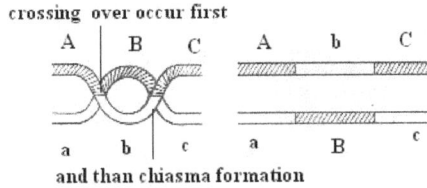

crossing over occur first

and than chiasma formation

Showing crossing over and chiasma formation - Chiasmatype theory

2. Chiasmatype theory or one plane theory. This theory was proposed by F.A. Janssens (1909) breakage occurs first, and the broken strands then reunite. Chiasmata are thus evidence, but not the causes, of a crossovers.

Recombination During Meiosis:- John Belling suggested that no break was necessary and proposed the copy choice model. He believed that crossing over might occur during duplication of homologous chromosomes and might brought about due to novel attachments formed between newly synthesized genes. While studying meiosis in some plant species. He visualized genes as beads (described as chromomeric), connected by non-genic inter chromomeric regions. The newly synthesized daughter chromatids is derived due to copying of one chromosomes upto certain region and then switching on to the other homologous chromosome for copying the remaining portion or region of the chromosomes. The new chromatid would have a new arrangement, but no breaks and rejoining need be involved. This was such an attractive idea that the hypothesis in some form held center stage for nearly thirty years.

New chromosome arrangements were associated with recombinant genes, using chromosomes marked by mutant genes and morphological differences at each end demonstrated in Drosophila melanogaster and Zea mays. That crossing over was correlated with segmental interchange between homologous chromosomes.

DNA models during 1960s,DNA models had become very popular, the widely accepted model for DNA crossover was first proposed by Robin Holliday.

Homologous recombination during meiosis has following important features.

Two homologous DNA molecules line up (e.g., two nonsister chromatids line up during meiosis).

Cuts in one strand of both DNAs, a double-strand break in a DNA molecule is enlarged by an exonuclease, such that the single-strand extension with a free 3'-hudroxyl group is left at the broken end.

The exposed 3' ends invade the intact duplex DNA, and this is followed by branch migration and/or replication to create a pair of crossover structures,(The cut strands cross and join homologous strands)called Holliday junctions(Holliday structure).

Heteroduplex region is formed by branch migration, cleavage of the two crossover creates two complete recombinant products

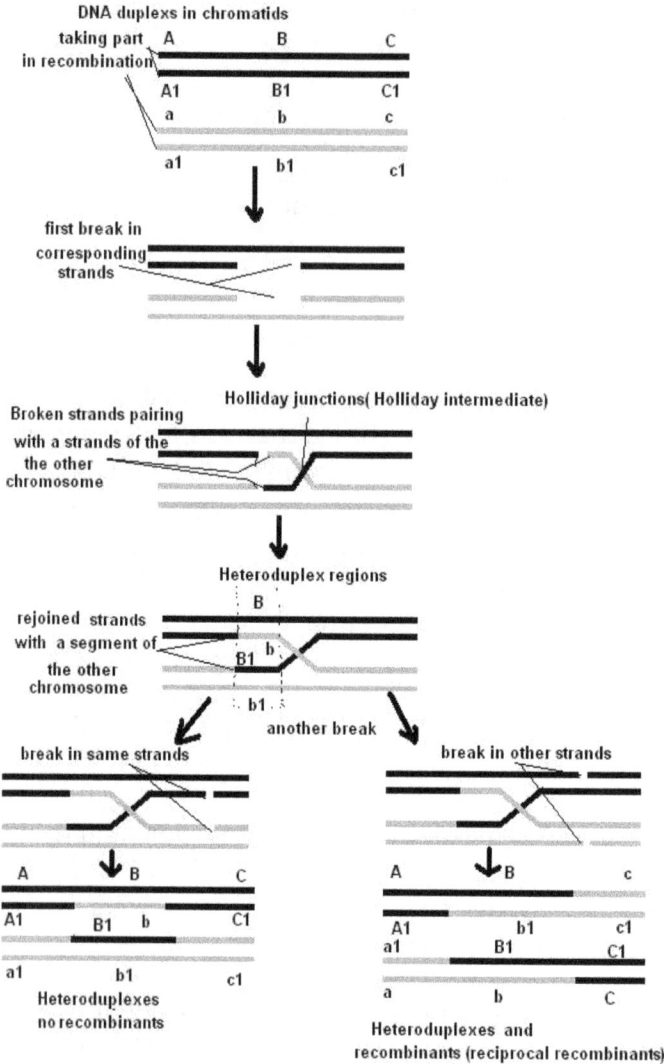

Holliday model of DNA cross over

In this double-strand break repair model for recombination, the 3' ends are used to initiate the genetic exchange. Once paired with the complementary strand on the intact homolog, a region of hybrid DNA is created containing complementary strands from two different parental DNA. Each of the 3' ends can then act as a primer for DNA replication.

The structure that are formed, called Holliday Intermediates are a feature of homologous genetic recombination pathways in all organisms. Homologous recombination can vary in many ways from one species to another, but most of the process are same.

DNA strands may be cut along either the vertical line or horizontal line and break the Holliday intermediate so that the two recombinant products carry genes in the same linear order as the original, unrecombined chromosomes.

If cleaved occurs on vertical line, the DNA flanking the region containing the hybrid DNA is not recombined; if cleaved occurs on the horizontal line, the flanking DNA is recombined.

Mutations

Mutation is defined as any sudden and drastic heritable change in gene which is not traceable or ascribable to segregation or recombination. According to Darwin sudden appearance of new hereditary character in the offspring of plants. Bateson said that mutation is a discontinuous variation. How even mutations play important role in evolution of a new species when a major set of characters is subjected to be mutated.

Mutations may be natural or induced and may be occur at chromosome level or at gene or molecular level or may takes place involving the cytoplasm or cytoplasmic organelles like plastids.

Classification of mutation

Plant geneticists have classified mutations in different ways for shake of their own convenience

Ammato classified mutation as:

- Gene mutation (mutation at gene level).

- Chromosomal mutation (structural level of chromosome).

- Genomatic mutation (Mutation in chromosome number).

According to Darlington and Mather mutations are following five types:

- Gene mutation.

- Structural mutation.

- Plastid mutation.

- Numerical mutation.

- Cytoplasmic mutation.

Recent studies on mutation in plants have been classified under following types:

Chromosomal Mutation

i)   Chromosomal aberrations (structural mutation).

ii) Geomantic or numerical mutations (Poly ploides).

iii) Germinal mutation.

iv) Somatic mutation.

v) Gene mutation or point mutation.

vi) Plastid mutation

vii) Cytoplasmic mutation.

Chromosomal mutation refers to any change in the structure or gross morphology of chromosome and change in chromosome number. The former case involve loss or gain or any alteration in chromosome which called chromosomal aberrations and the later is the change in basic chromosomal aberrations and the later is the change is basic chromosome number of a species and called polyploidy or numerical mutation.

I. Chromosomal Aberrations

These include structural changes which takes place during meiosis. These are following types:

- Deletion or deficiency: In this case if a chromosome broken in to pieces and reunion takes place without taking one or more pieces with loss of a segment. It has great cytological and genetic effect on organism and may be lethal.

- Duplication: If an extra chromosomal part is added to normal character, it is called duplication. Duplication takes place when deleted part of a chromosome added to another chromosome.

- Translocation: When a part of chromosome is broken into two or more pieces and some fragment is transferred from the broken chromatid and rejoin with the other chromatid then it is called translocation. It may be simple or reciprocal causing disturbance in position of genes. It may cause sterility in plants.

- Inversion: In this process a piece of chromosome is removed and reinserted in reverse order changing the gene sequences. It may be paracentric or pericentric type. It causes position effect resulting alteration in geneaction.

II. Poly Ploidy: (Numerical Mutation)

In this type of mutation basic genomic set changed due to disjunction or defective meiosis or nuclear division. Thus changed into 3n or 4n or 6n etc. these polyploidy can be classified into Autopolyploidy and allopolyploidy.

a) Autopolyploidy

When the basic chromosome set of an individual is multiplied due to union of two diploid gametes or somatic doubling of chromosomes or union of one haploid and one diploid gamets. It may be autotriploid, autotetraploid etc. these are found in avaena, coffee, apple, banana, sugarcane, chili etc.

(b) Allopolyloids

Allopolyploids formed by multiplication of chromosome sets of a hybrid of two diploid species. Suppose in a cross between a species – X (AA) and species Y (BB) is made it will give rise to F1 hybrid with chromosome doubling and a alloteraploid having (A,A,B,B) genomic set will be obtained.

Allopolyploidy can be induced and have played important role in evolution of new species.

Example: Raphanobrassica, Gossipium hissutum etc.

III. Germinal Mutation

Those mutations occur in the germplasm of an organism and may occur at any time during life cycle are called germinal mutation. The effect of these mutations expressed in the progenies. This mutation is found in wheat as dwarfism. Expression of germinal Mutation depends upon whether it is of dominant or recessive form. The progeny produced will be mutant if the gamete posses mutant gene.

IV. Somatic Mutation

Somatic mutations are the mutation occurs in the somatic cells of an individual, which cannot be passed to the offspring through the gamets. This mutated genes lost with the death of the individual.

In somatic mutations, the extent of phenotypic effect depends on various factors, like the stage of life cycle. It cannot be pass on to the germ cells and only inherited to the next generation of progeny, which produced from the mutated somatic cell.

This mutation cause cancerous growth and defected metabolism of cells and tissues produced by the mutated cell. It is found in vegitatively and asexulally reproducing plants and animals.

Such mutations used in production of delicious apples, navel orange and many other fruits and flower production.

Now it is commercially used in horticultural practices.

V. Gene Mutation or Point Mutation

These gene mutations are intragenic in which alteration in the structure of DNA molecule within a gene occurs. These occurs change in the normal base sequence of DNA molecule which leads to modification of structural characteristic or enzymatic activity of an individual. The unit of gene mutation, i.e. a specific nucleotide or nucleotides called muton.

In this type of mutation tautomerism occur in the structure of base molecule, which enable it for unusual pairing. "Adenine instead of bonding with "Thymine" binds with guanine or other bases or other tautomeric forms of bases.

Gene mutations may be spontaneous or may be induced by application of mutagenic agents.

Gene mutations may occur under following types on the basis of their mode of occurrence.

    i) Frame shift mutations

    ii) Deletion mutations

    iii) Insertion mutations

    iv) Substitution mutations

    v) Transition mutation

    vi) Transition mutation

    vii) Transversion mutation

    viii) Inversion mutation

    ix) Missense mutation (Formation of codon carried other amino acid in sequence)

    x) Nonsence mutation (Forming nonsence codons at the placed of normal codons)

    xi) Silent mutations (Point mutation having no phenotypic expression)

VI. Plastid Mutation

Mutations in the genetic materials in plastids are called plastid mutation and it is governed by the self duplicating non-med3elian genetic material called plastogenes. Plastogenic mutation cause defective plastid characters which traced in maize, barley and rice.

VII. Cytoplasmic Mutation

Mutation, which takes place in the nucleus free or extra chromosomal genetic material is called cytoplasmic mutation. These mutations effect cytoplasmic inheritance chlorophyl deficiency in algae, like chlamydomonas, enzmatic alteration in yeast and antibiotic resistant in bacteria are the consequences of cytoplasmic mutation. These mutations occure at plasmogenes or cytogenes or plasmones or plasmid.

Role of Mutation in Plant Breeding

Plant breeding aims at improving the crop quality but improving the heredity through the cross hybridization technique. In plants mutations can be artificially incused by mutagenic agents and there utilization for production of new superior varieties of species from traditional variety. This process is called mutation breeding.

The history of mutation breeding in India started in 1935 at Bose research institute, Calcutta and established at IARI, New Delhi.

a) Mutation breeding in wheat (Triticum species)

By the application of Eradication and chemical mutagens mutation being used to introduced in wheat varieties. By this process the resistance variety NP836 was derived from NP 799 at India agricultural research institute (IARI), New Delhi. This is done by application gamma ray on NP 799 variety of wheat. Thus NP 836 is a mutant variety.

b) Mutation breeding in rice (Oryzaesativa)

Mutation breeding in rice is very common in south and south-east Asia. In rice certain chemical mutafgenens has been used to produce polyploid varieties of rice and hybridised with the diploids producing high yielding and resistant varieties.

The high yielding varieties of rice produced by mutation breeding is P 500.28. This variety is obtained from the T-1145 variety at Bose's institute, Calcutta.

c) Mutation breeding in cotton (Gossypium)

Mutation beading achieved evolving improved variety in cotton. A caiety named indore-2 was developed from "Malwa Upland -4" by X-ray treatment.

Mutations in thevariey mescilla cabala by X-ray treatment result increase in 40-50% in fiber production.

M.A.2 H-190. Indore-2 L.SS Bury-0394, 320-F and H-14 are mutation improved varieties of cotton.

d) Mutation breeding in sugarcane

Both eradication and chemical mutagens are used to induce maturation in sugarcane. In sugarcane nodal buds are exposed to radiation in field and the mutant buds or tillers are selected in F1 and F2 generations through artificial crosses.

Some popular higher quality sugarcane varieties evolved through mutation breeding are H.M.658, H.M. -661, Co-213, Co-602, Co-612 etc.

e) Mutation breeding in potato

Mutations also introduce in potato crop through mutation breeding. It aims at the production of early harvesting varieties and high yielding variety.

These includes eradication and chemical mutagenic products and through cross breeding. This is done in flowering or by exposing the seed tubers.

Gene Interaction

Mendelian genetics does not explain all kinds of inheritance for which the phenotypic ratios in some cases are different from Mendelian ratios (3:1 for monohybrid, 9:3:3:1 for di-hybrid in F2). This is because sometimes a particular allele may be partially or equally dominant to the other or due to existence of more than two alleles or due to lethal alleles. These kinds of genetic interactions between the alleles of a single gene are referred to as allelic or intra- allelic interactions.

Type	Ratio	Interaction	Example
A. Allelic interactions			
1. Incomplete dominance	1:2:1	Partial dominance.	Flower color in snapdragon.
(a) Monohybrid	1:2:1:2:4:2:1:2:1	Partial dominance at both the gene pairs.	Human blood group (ABO and MN).
	3:6:3:1:2:1	Complete dominance at one gene pair and partial dominance at the other.	Cattle (horn and hair color).
2. Lethal factor	2:1/3:0	Homozygous condition causes death.	Yellow coat color in mice, albino seedling in barley.
3. Multiple alleles	-	Occurrence of more than two alleles in a single locus.	ABO blood group system in human, self-sterility in tobacco.
B. Non-allelic Interactions			
4.Simple Interaction	9:3:3:1	New phenotypes resulting from interaction between two dominants and also between two recessives.	Comb types in fowl, streptocarpus flower color.
5.Complementary factor	9:7	Two dominant genes are complementary to each other in their effect.	Flower color in sweet pea.

6. Epistasis			
(a) Recessive	9:3:4	A homozygous recessive gene is epistatic to other gene.	Coat color in mice, grain color in maize.
(b) Dominant	12:3:1	A dominant gene inhibits to other geges	Fruit color in summer squash.
7. (a) Inhibitory factor	13:3	One dominant gene partially inhibits the expression of the other.	Leaf color in rice.
(b) Inhibitory factor with partial dominance.	7:6:3	One dominant gene partially inhibits the expression of the other.	Hair direction in guineapig.
8. Polymorphic gene	9:6:1	New phenotype from interaction between two dominant genes.	Awn length in barley.
9. (a) Duplicate gene	15:1	Dominant allele of either gene pair, alone or together, are similar in phenotypic effect.	Capsule shape in shepherd's purse.
(b) Duplicate gene with Dominance modification.	11:5	Dominance due to two non-allelic or allelic dominent alleles.	Pigment glands in cotton.
10. Multiple factors			
(a) Two loci	1:4:6:4:1	A quantitative trait controlled by several genes having cumulative effect.	Kernel color in wheat, skin color in human.
(b) Three loci	1:6:15:20:15:6:1		

Non-allelic or inter-allelic interactions also occur where the development of single character is due to two or more genes affecting the expression of each other in various ways.

Thus, the expression of gene is not independent of each other and dependent on the presence or absence of other gene or genes; These kinds of deviations from Mendelian one gene-one trait concept is known as Factor Hypothesis or Interaction of Genes.

Allelic Gene Interactions

Incomplete Dominance or Blending Inheritance (1:2:1)

A dominant allele may not completely suppress other allele, hence a heterozygote is phenotypically distinguishable (intermediate phenotype) from either homozygotes.

In snapdragon and Mirabilis jalapa, the cross between pure bred red-flowered and white-flowered plants yields pink-flowered F_1 hybrid plants (deviation from parental phenotypes), i.e., intermediate of the two parents. When F_1 plants are self-fertilized, the F_2 progeny shows three classes of plants in the ratio 1 red: 2 pink: 1 white instead of 3:1.

Parents	Red flower × White flower
	RR rr
Gametes	Ⓡ ⓡ

F_1 — Rr Pink Flower

Selfing

F_2

♀ \ ♂	R	r
R	RR Red	Rr Pink
r	Rr Pink	rr White

Genotype	Phenotype	Ratio
RR	Red	1
Rr	Pink	2
rr	White	1

Inheritance of flower color in snapdregon

Therefore, a F_1 di-hybrid showing incomplete dominance for both the characters will segregate in F_2 into (1:2:1) X (1:2:1) = 1:2:1:2:4:2:1:2:1. And a F_1 di-hybrid showing complete dominance for one trait and incomplete dominance of another trait will segregate in F_2 into (3:1) x (1:2:1) = 3:6:3:1:2:1.

Multiple Alleles

A gene for particular character may have more than two allelomorphs or alleles occupying same locus of the chromosome (only two of them present in a diploid organism). These allelomorphs make a series of multiple alleles.

Self-sterility in tobacco is determined by the gene with many different allelic forms. If there are only three alleles (s_1, s_2, s_3), the possible genotypes of plants are s_1s_2, s_1s_3, s_2s_3 (always heterozygous), homozygous genotypes (s_1s_1, s_2s_2, s_3s_3) are not possible in a self-incompatible species.

In this case, pollen carrying an allele different from the two alleles present in the female plant will be able to function resulting in restriction of fertility.

Non Allelic Gene Interactions

Simple Interaction (9:3:3:1)

In this case, two non-alleiic gene pairs affect the same character. The dominant allele of each of the two factors produces separate phenotypes when they are alone. When both the dominant alleles are present together, they produce a distinct new phenotype. The absence of both the dominant alleles gives rise to yet another phenotype.

One of the best example of simple interaction is Streptocarpus flower color.

Complementary Factor (9:7)

Certain characters are produced by the interaction between two or more genes occupying different loci inherited from different parents. These genes are complementary to one another, i.e., if present alone they remain unexpressed, only when they are brought together through suitable crossing will express.

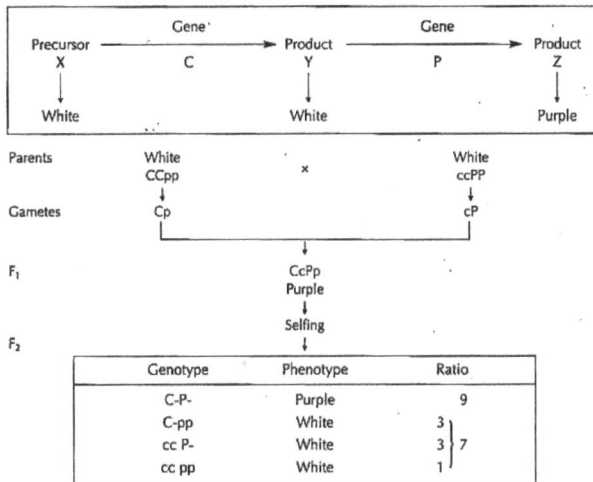

Inheritance of flower color in lathyrus odoratus

In sweet pea (Lathyrus odoratus), both the genes C and P are required to synthesize anthocyanin pigment causing purple color. But absence of any one cannot produce anthocyanin causing white flower. So C and P are complementary to each other for anthocyanin formation.

Involvement of more than two complementary genes is possible, e.g., three complementary genes governing aleurone color in maize.

Epistasis

When a gene or gene pair masks or prevents the expression of other non-allelic gene, called epistasis. The gene which produces the effect called epistatic gene and the gene whose expression is suppressed called hypostatic gene.

(a) Recessive Epistasis or Supplementary Factor (9:3:4)

In this case, homozygous recessive condition of a gene determines the phenotype irrespective of the alleles of other gene pairs, i.e., recessive allele hides the effect of the other gene.

The grain color in maize is governed by two genes — R (red) and Pr (purple). The recessive allele rr is epistatic to gene Pr.

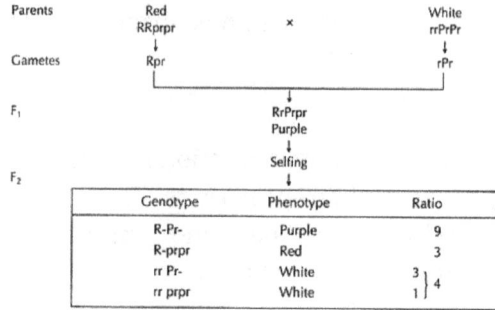

Parents	Red	×	White
	RRprpr		rrPrPr
	↓		↓
Gametes	Rpr		rPr

F₁ — RrPrpr Purple ↓ Selfing

F₂

Genotype	Phenotype	Ratio
R-Pr-	Purple	9
R-prpr	Red	3
rr Pr-	White	3 ⎫ 4
rr prpr	White	1 ⎭

Inheritance of grain color in maize

(b) Dominant Epistasis (12:3:1)

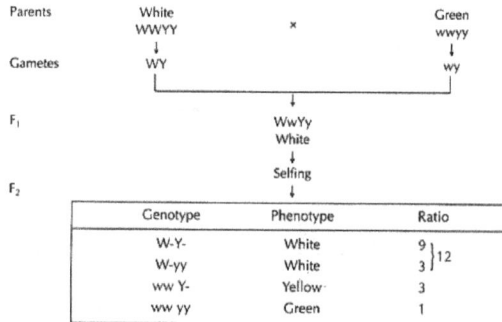

Parents	White	×	Green
	WWYY		wwyy
	↓		↓
Gametes	WY		wy

F₁ — WwYy White ↓ Selfing

F₂

Genotype	Phenotype	Ratio
W-Y-	White	9 ⎫ 12
W-yy	White	3 ⎭
ww Y-	Yellow	3
ww yy	Green	1

Inheritance of fruit color in summer squash

Sometimes a dominant gene does not allow the expression of other non-allelic gene called dominant epistasis. In summer squash, the fruit color is governed by two genes. The dominant gene W for white color, suppresses the expression of the gene Y, which controls yellow color. So yellow color appears only in absence of W. Thus W is epistatic to Y. In absence of both W and Y, green color develops.

Inhibitory Factor

Inhibitory factor is such a gene which itself has no phenotypic effect but inhibits the expression of another non-allelic gene; in rice, purple leaf color is due to gene P, and p causing green color. Another non-allelic dominant gene I inhibits the expression of P but is ineffective in recessive form (ii). Thus the factor I has no visible effect of its own but inhibits the color expression of P.

Polymorphic Gene (9:6:1)

Here two non-allelilc genes controlling a character produce identical phenotype when they are alone, but when both the genes are present together their phenotypes effect is enhanced due to cumulative effect. In barley, two genes A and B affect the length of awns.

Gene A or B alone gives rise to awns of medium length (the effect of A is same as B); but when both present, long awn is produced; absence of both results aweless.

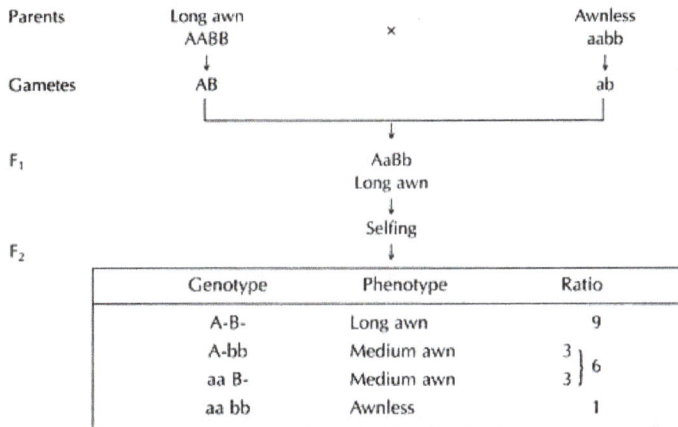

Parents	Long awn AABB	×	Awnless aabb
Gametes	AB		ab

F₁ : AaBb Long awn → Selfing

F₂ :

Genotype	Phenotype	Ratio
A-B-	Long awn	9
A-bb	Medium awn	3 } 6
aa B-	Medium awn	3
aa bb	Awnless	1

Inheritance of awns in barley

Duplicate Gene (15:1)

Sometimes a character is controlled by two non-allelic genes whose dominant alleles produce the same phenotype whether they are alone or together. In Shepherd's purse (Capsella bursa-pastoris), the presence of either gene A or gene B or both results in triangular capsules; when both these genes are in recessive forms, the oval capsules produced.

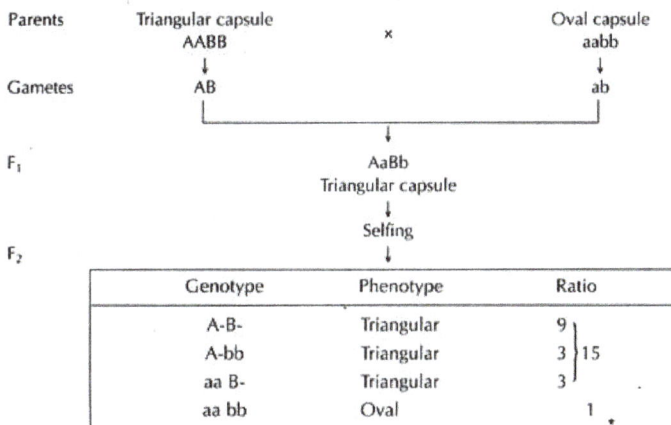

Parents	Triangular capsule AABB	×	Oval capsule aabb
Gametes	AB		ab

F₁ : AaBb Triangular capsule → Selfing

F₂ :

Genotype	Phenotype	Ratio
A-B-	Triangular	9
A-bb	Triangular	3 } 15
aa B-	Triangular	3
aa bb	Oval	1

Inheritance of fcapsule shape in shepherd's purse

Duplicate Gene with Dominance Modification (11:5)

A character controlled by two gene pairs showing dominance only if two dominant alleles are present. Dominant phenotype will thus be produced only when two non-allelic dominant alleles or two allelic dominant alleles are present. Such a case is found in pigment glands of cotton.

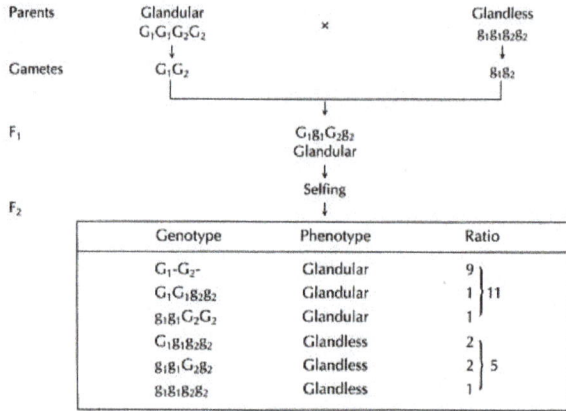

Parents	Glandular $G_1G_1G_2G_2$	×	Glandless $g_1g_1g_2g_2$
Gametes	G_1G_2		g_1g_2

F$_1$: $G_1g_1G_2g_2$ Glandular

Selfing

F$_2$:

Genotype	Phenotype	Ratio
G_1-G_2-	Glandular	9
$G_1G_1g_2g_2$	Glandular	1 } 11
$g_1g_1G_2G_2$	Glandular	1
$G_1g_1g_2g_2$	Glandless	2
$g_1g_1G_2g_2$	Glandless	2 } 5
$g_1g_1g_2g_2$	Glandless	1

Inheritance of pigment glands in cotton

Multiple Factors and Polygenic Gene Inheritance

Though some characters (qualitative) show discontinuous variation but a majority of characters (quantitative, e.g., height, weight, etc.) exhibit continuous variation. Yule, Nilsson-Ehle, East suggested that quantitative variation is controlled by large number of individual genes called polygenic systems and the inheritance could be explained on the basis of multiple factor hypothesis.

The hypothesis states that for a given quantitative trait there could be several genes, which are independent in their segregation and had cumulative effect on the phenotype.

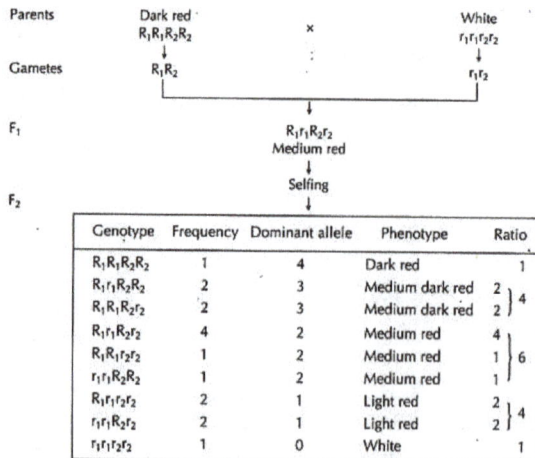

Parents	Dark red $R_1R_1R_2R_2$	×	White $r_1r_1r_2r_2$
Gametes	R_1R_2	:	r_1r_2

F$_1$: $R_1r_1R_2r_2$ Medium red

Selfing

F$_2$:

Genotype	Frequency	Dominant allele	Phenotype	Ratio
$R_1R_1R_2R_2$	1	4	Dark red	1
$R_1r_1R_2R_2$	2	3	Medium dark red	2
$R_1R_1R_2r_2$	2	3	Medium dark red	2 } 4
$R_1r_1R_2r_2$	4	2	Medium red	4
$R_1R_1r_2r_2$	1	2	Medium red	1 } 6
$r_1r_1R_2R_2$	1	2	Medium red	1
$R_1r_1r_2r_2$	2	1	Light red	2
$r_1r_1R_2r_2$	2	1	Light red	2 } 4
$r_1r_1r_2r_2$	1	0	White	1

Inheritance of kernel color in wheat

Kernel color in wheat is a quantitative character and controlled by two different genes. The heterozygote is intermediate in color between the two homozygotes. Both the dominant genes have small and equal (or almost equal) effects on seed color. F_1 heterozygote for two genes will segregate in F_2 in the ratio 1:4: 6:4:1.

The intensity of seed color depends on the number of dominant alleles present, i.e., their effects are cumulative in nature. It is now known that there are three genes involved in kernel color in wheat, thus a F, heterozygous for all three genes will segregate in F_1 in the ratio 1 : 6 : 15 : 20 : 15 : 6 : 1.

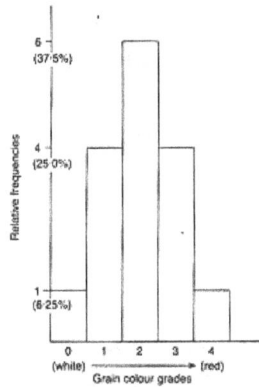

Relative frequencies of different grain color in wheat in F_2 population
derived from two strains differing in two gene pairs

Corolla length in Nicotiana is another quantitative character. The crosses made between two inbred varieties of N. longiflora or N. tabaccum (tobacco) differing in corolla length, show F_1 with uniform corolla length but F_2 exhibits a greater degree of variation.

Mean value of F_3 derived from single F_2 plant with particular corolla length differs greatly from other single plant F_3 progenies. It is thus obvious that F_2 plants differ genetically.

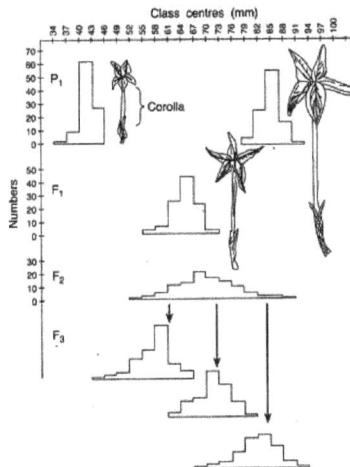

Frequancy distribution of lengths of corolla in two different varieties of nicotiana used
as parents and F_1, F_2 and F_3 generations derived from them

Pleiotropy

Pleiotropy refers to the phenomenon of a single gene affecting multiple traits. Pleiotropy implies a mapping from one thing at the genetic level to multiple things at a phenotypic level. The natures of the things differ in different contexts. The term pleiotropy is divided into three broad classes. The classifications are not mutually exclusive—indeed, they typically overlap—but they do represent different and potentially contradictory perspectives.

In *Molecular-Gene Pleiotropy*, the question is about the number of functions a molecular gene has. These functions can be defined genetically, by the number of measured traits affected by a knockout, but also biochemically, for example, by the number of protein-protein interactors a gene has or the number of reactions it catalyzes.

In *Developmental Pleiotropy*, mutations –not molecular genes –are the relevant units. Developmental pleiotropy is a feature of the genotype-phenotype map that defines the genetic and evolutionary autonomy of aspects of phenotype, independent of fitness. This is the mutational pleiotropy underlying the diverse manifestations of syndromic diseases, the ontogenetic pleiotropy that underlies classical questions about allometry and heterochrony, and molecular pleiotropy that underlies questions about relative importance of cis-regulatory vs. protein-coding variants.

In *Selectional Pleiotropy*, the question is about the number of separate components of fitness a mutation affects. In some cases, the multiple fitness components are life-history traits of a single individual, which is at the heart of the antagonistic pleiotropy model for the evolution of aging. In other cases, the mutational effects are manifest in different individuals in a population, which is the basis for sexually antagonistic pleiotropy and pleiotropic trade-offs underlying local adaptation. A key feature of selectional pleiotropy is that traits are defined by the action of selection and not by the intrinsic attributes of the organism.

Penetrance and Expressivity

Penetrance refers to the probability of a gene or trait being expressed. In some cases, despite the presence of a dominant allele, a phenotype may not be present. One example of this is polydactyly in humans (extra fingers and/or toes). A dominant allele produces polydactyly in humans but not all humans with the allele display the extra digits. "Complete" penetrance means the gene or genes for a trait are expressed in all the population who have the genes. "Incomplete" or 'reduced' penetrance means the genetic trait is expressed in only part of the population. The penetrance of expression may also change in different age groups of a population. Reduced penetrance probably results from a combination of genetic, environmental, and lifestyle factors, many of which are unknown. This phenomenon can make it challenging for genetics professionals to interpret a person's family medical history and predict the risk of passing a genetic condition to future generations.

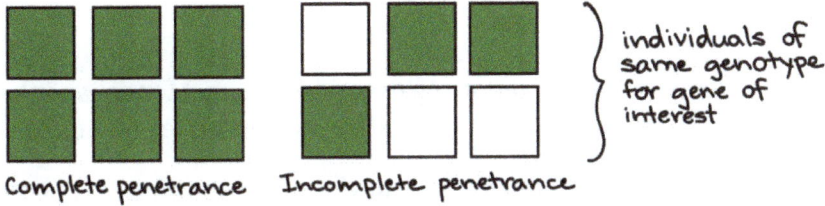

Complete penetrance   Incomplete penetrance

individuals of same genotype for gene of interest

Expressivity on the other hand refers to variation in phenotypic expression when an allele is penetrant. Back to the polydactyly example, an extra digit may occur on one or more appendages. The digit can be full size or just a stub. Hence, this allele has reduced penetrance as well as variable expressivity. Variable expressivity refers to the range of signs and symptoms that can occur in different people with the same genetic condition. As with reduced penetrance, variable expressivity is probably caused by a combination of genetic, environmental, and lifestyle factors, most of which have not been identified. If a genetic condition has highly variable signs and symptoms, it may be challenging to diagnose.

Narrow expressivity   Variable expressivity

individuals of same genotype for gene of interest

Cytoplasmic Inheritance

Cytoplasmic inheritance is also called Extranuclear inheritance. Extranuclear inheritance is a non-Mendelian form of heredity that involves genetic information located in cytoplasmic organelles, such as mitochondria and chloroplasts, rather than on the chromosomes found in the cell nucleus.

Extranuclear genes, also known as cytoplasmic genes, are located in mitochondria and chloroplasts of a cell rather than in the cell's nucleus on the chromosomes. Both egg and sperm contribute equally to the inheritance of nuclear genes, but extranuclear genes are more likely to be transmitted through the maternal line because the egg is rich in the cytoplasmic organelles where these genes are located, whereas the sperm contributes only its nucleus to the fertilized egg.

Therefore, extranuclear genes do not follow genetic pioneer Gregor Mendel's statistical laws of segregation and recombination. Cytoplasmic genes are of interest in understanding evolution, genetic diseases, and the relationship between genetics and embryology.

Chloroplasts

As early as 1909, geneticists were reporting examples of non-Mendelian inheritance in higher plants, usually green and white variegated patterns on leaves and stems.

These patterns seemed to be related to the behavior of the chloroplasts, photosynthetic organelles in green plants. Because of the relatively large size of chloroplasts, scientists have been able to study their behavior in dividing cells with the light microscope since the 1880's.

Like mitochondria, chloroplasts contain their own deoxyribonucleic acid (DNA) and ribonucleic acid (RNA). Although chloroplast DNA (cpDNA) contains many of the genes needed for chloroplast function, chloroplasts do not seem to be totally autonomous; nuclear genes are required for some chloroplast functions.

Another interesting case of extranuclear inheritance in plants is that of cytoplasmic pollen sterility. Many species of plants seem to produce strains with cytoplasmically inherited pollen sterility.

Advances in experimental methods made in the 1960's allowed scientists to demonstrate that organelles located in the cytoplasm contain DNA. This finding came as a great surprise to most biologists.

In 1966 the first vertebrate mitochondrial DNA (mtDNA) was isolated and characterized. Like bacterial DNA, mtDNA generally consists of a single double helix of "naked," circular DNA. The mitochondrial genome is usually smaller than that of even the simplest bacterium.

Most of the proteins in the mitochondrion are encoded by nuclear genes, but mtDNA contains genes for mitochondrial ribosomal RNAs, transfer RNAs, and some of the proteins of the electron transport system of the inner membrane of the mitochondrion.

Extranuclear DNA

The DNA found in chloroplasts and mitochondria is chemically distinct from the DNA in the nucleus. Moreover, the extranuclear genetic systems behave differently from those within the nucleus.

Even more surprising is the finding that mitochondria have their own, slightly different version of the genetic code, which was previously thought to be common to all organisms, from viruses to humans. In general, because of its greatly smaller size, the DNA found in cytoplasmic organelles has a limited coding capacity.

Extranuclear DNA

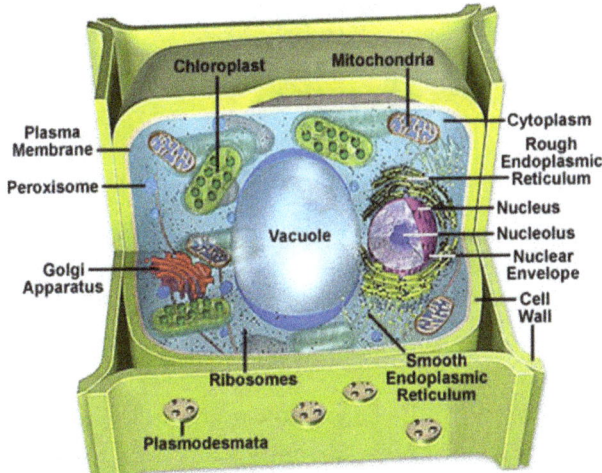

Thus, by identifying the functions under the control of mitochondrial or chloroplast genes, all other functions carried on by the organelle can be assigned to the nuclear genome. Coordinating the contributions of the organelle and the nuclear genomes is undoubtedly a complex process.

In addition to the genes found in mitochondria and chloroplasts, extranuclear factors are found in various kinds of endosymbionts (symbiotic organisms that live within the cells of other organisms) and bacterial plasmids. Some biologists think that all organelles may have evolved from ancient symbiotic relationships. Endosymbionts may be bacteria, algae, fungi, protists, or viruses.

Unlike the mitochondria and chloroplasts, some endosymbionts seem to have retained independent genetic systems. The "killer" particles in paramecia, discovered by T.M. Sonneborn in the 1930's, provide a historically significant example. After many years of controversy, the killer particles were identified as bacterial symbionts.

These cytoplasmic entities are not vital to the host cell, as the paramecia are capable of living and reproducing without them. Certain peculiar non-Mendelian conditions found in fruit flies also appear to be caused by endosymbionts.

Although bacteria lack nuclei, their circular DNA is usually referred to as bacterial chromosomes. Some bacteria also contain separate DNA circles smaller than the bacterial chromosome. In the 1950's Joshua Lederberg proposed the name "plasmid" for such extrachromosomal hereditary determinants.

One of the earliest and best known examples of cytoplasmic inheritance is that discovered by Correns in a variegated variety of the four-o'clock plant *Mirabilis jalapa*.

Variegated plants have some branches which carry normal green leaves, some branches with variegated leaves (mosaic of green and white patches) and some branches which have all white leaves.

Leaf variegation in mirabilis jalapa

Correns discovered that seed produced by flowers carried on the green branches gave progeny which were all normal green. It made no difference whether the phenotype of the branch which carried the flower used for pollen was green, white or variegated. Seed taken from white branches likewise gave all white progeny, regardless of the pollen donor phenotype. These of course died in the seedling stage. Seeds from flowers on variegated branches gave three kinds of progeny, green, white and variegated, in varying proportions; again regardless of the pollen donor phenotype. In other words, the phenotype of the progeny always resembled the female parent and the male made no contribution at all to the character. The effect is seen quite clearly in the difference which Correns found between reciprocal crosses:

$$♀ \text{ green } \times \text{ white } ♂ \rightarrow \text{ green progeny}$$

$$♀ \text{ white } \times \text{ green } ♂ \rightarrow \text{ white progeny}$$

The explanation for this unusual pattern of inheritance is that the genes concerned are located in the chloroplasts within the cytoplasm, not in the nucleus, and are therefore transmitted only through the female parent. In eukaryote organisms the zygote normally receives the bulk of its cytoplasm from the egg cell and the male gamete contributes little more than a nucleus. Any genes contained in the cell organelles of the cytoplasm will therefore show maternal inheritance. The leaf variegation is due to two kinds of chloroplasts: normal green ones and defective ones lacking in chlorophyll pigment. Chloroplasts are genetically autonomous (i.e. self-determining) and have their own system of heredity in the form of chloroplast 'chromosomes'. These are small circular naked DNA molecules which carry genes controlling *some* aspects of chloroplast structure and function. A mutation in one of these genes, which affects the synthesis of chlorophyll as in *Mirabilis*, will therefore follow the chloroplast in its transmission and will not be inherited in the same way as a nuclear gene.

In figure inheritance of leaf variegation in *Mirabilis jalapa*. The character is controlled by cytoplasmic genes located in the chloroplast 'chromosome'. Chloroplasts are self-perpetuating cell organelles and during sexual reproduction they are only transmitted through the cytoplasm of the egg cell, as undifferentiated protoplastids. They are not inherited through the pollen. The progeny of crosses therefore have the characters of the female parent and show maternal inheritance.

The other important point to note about the inheritance of chloroplasts is that they have no regular means of distribution, such as chromosomes do at mitosis, where they can be equally shared out to the daughter cells following division. A plant that begins life as a zygote containing a mixture of normal and mutant chloroplasts cannot therefore maintain the same mixture in all of its somatic cells. The two kinds of plastids are shared out randomly during cell division, according to the way they happen to be placed in the cytoplasm when it is partitioned. Some branches of variegated plants may therefore remain mosaic while others, by chance, may turn out to contain all white or all green chloroplasts in all of their cells. In a similar way the flowers on variegated branch may be of three kinds. Some will have egg cells with all green chloroplasts, some egg cells with all white and others will retain a mixture.

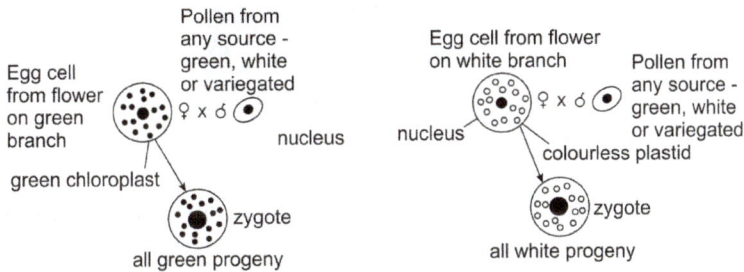

Chloroplast Genome

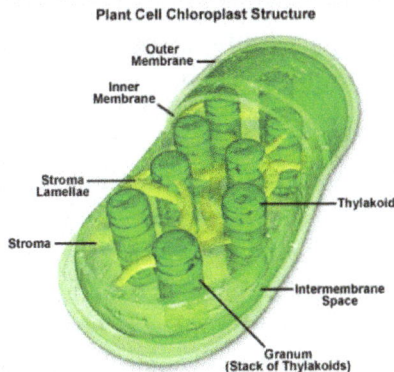

It is now known that the chloroplasts of plants carry their genetic information in the form of small circular DNA molecules, similar in size and form to the chromosomes of bacteria. These DNA molecules contain genes which code for some of the proteins and RNAs used in chloroplast structure and function; and it is mutations in these genes which are most likely to be responsible for the leaf variegation effects described above. It must also be emphasised that chloroplasts are not totally independent of the nucleus in their heredity; most of their proteins are coded by nuclear genes, and mutations in these show normal Mendelian patterns of inheritance.

The DNA molecules which make up the chloroplast genome are 'naked' ones and bear no resemblance to the chromosomes of the nucleus, which are much larger and are composed of both protein and DNA. The really surprising thing about the chloroplast DNA is the large number of copies which are present: up to 300 in a mature plastid. Since an average of 160 chloroplasts are present in a mesophyll cell of the mature leaf of a cereal such as wheat, this means that there may be as many as 48 000 chloroplast 'chromosomes' per mesophyll cell. The reason for this enormous redundancy of genetic information is unknown.

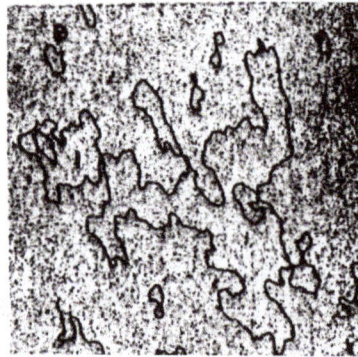

Figure: Electron micrograph of a single circular molecule of chloroplast DNA (ctDNA) from the lettuce plant. The chloroplast 'chromosome' has a length of 155 000 base pairs of DNA. The small circles are 'chromosomes' of the virus fX174 (5370 nucleotides of DNA) which are included as a standard for calculating the size of the chloroplast 'chromosome'.

Other Examples of Cytoplasmic Inheritance

Leaf variegation due to chloroplast mutation is known in numerous other genera of plants: *Epilobium* and *Pelargonium* are two examples.

Many of the other examples of cytoplasmic inheritance, in a variety of species, appear to involve characters which are associated with functions of the mitochondria. They have to do with defects in growth and ATP energy metabolism. Well known cases include the 'Poky' (slow growing) mutants in the fungus *Neurospora* and 'Petite' mutants' in brewers yeast. The mitochondria, like the chloroplasts, are self-replicating organelles which contain their own genes and have a limited number of characters which are independent of the nucleus. They are transmitted mainly through the female line and mutations in their genes show the same pattern or maternal inheritance. Mitochondrial 'chromosomes' have a similar circular configuration of 'naked' DNA as chloroplasts. In a typical haploid yeast cell each of the mitochondria contains in the region of 50 small circular 'chromosomes'.

Hybridization

Individual produced as a result of cross between two genetically different parents is known as hybrid. The natural or artificial process that results in the formation of hybrid is known as hybridization.

The production of a hybrid by crossing two individuals of unlike genetical constitution is known as hybridization. Hybridization is an important method of combining characters of different plants. Hybridization does not change genetic contents of organisms but it produces new combination of genes.

The first natural hybridization was recorded by Cotton Mather in corn. The first artificial interspecific plant hybrid was produced by Thomas Fairchild in 1717. It is commonly known as 'Fairchild Mule'.

Hybridization was first of all practically utilized in crop improvement by German botanist Joseph Koerauter in 1760. Mendel onward, the hybridization had become the key method of crop improvement. Today, it is the most common method of crop improvement, and the vast majority of crop varieties have resulted from hybridization.

Objectives of Hybridization

   I.   To artificially create a variable population for the selection of types with desired combination of characters;

   II. To combine the desired characters into a single individual, and

   III. To exploit and utilize the hybrid varieties.

Types of Hybridization

Hybridization may be of following types:

   i) Intra-varietal hybridization:

     The crosses are made between the plants of the same variety.

   ii) Inter-varietal or Intraspecific hybridization:

     The crosses are made between the plants belonging to two different varieties.

   iii) Inter-specific hybridization or intra-genric hybridization:

     The crosses are made between two different species of the same genus.

   iv) Introgressive hybridization:

     Transfer of some genes from one species into the genome of the other species is known as introgressive hybridization. The crosses between different species of the same genus or different genera of the same family are also known as distant hybridization or wide crossing. Such crosses are called distant crosses.

Procedure of Hybridization

It involves the following steps:

  i) Selection of parents.

 ii) Selfing of parents or artificial self-pollination.

iii) Emasculation.

iv) Bagging

 v) Tagging

vi) Crossing

vii) Harvesting and storing the F, seeds

viii) Raising the F_1 generation.

Selection of Parents

The selection of parents depends upon the aims and objectives of breeding. Parental plants must be selected from the local areas and are supposed to be the best suited to the existing conditions.

Selfing of Parents or Artificial Self-pollination

It is essential for inducing homozygosity for eliminating the undesirable characters and obtaining inbreeds.

Emasculation

It is the third step in hybridization. Inbreeds are grown under normal conditions and are emasculated. Emasculation is the removal of stamens from female parent before they burst and shed their pollens.

It can be defined as the removal of stamens or anthers or the killing of the pollen grains of a flower without affecting in any way the female reproductive organs. Emasculation is not required in unisexual plants but it is essential in bisexual or self-pollinated plants.

Various methods used for emasculation are:

(a) Hand Emasculation or Forceps or Scissor Method

This method is generally used in those plants, which have large flowers. In this method the corolla of the selected flowers is opened and the anthers carefully removed with the help of fine-tip forceps.

Following are the important precautions while performing this method:

    i. Flowers should be selected at proper stage.

    ii. Stigma should be receptive and anthers should not have dehisced.

    iii. All the anthers should be removed from the flowers without breaking.

    iv. Stigma and ovary of the flower should not be damaged.

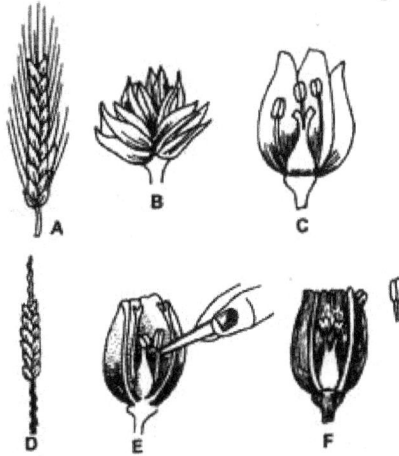

(A-F) Emasculation in wheat. (A) Spike of spikelets. (B) spikelet, (C) Floret, (D) Upper and lower spikelets removed, awns removed, upper portion of florets cut, (E) Anthers removed with the help of fine-tip forcept. (F) Removed anther.

(b) Hot Water Treatment

Removal of stamens with the help of forceps is very difficult in minute flowers. In such small hermaphrodite flowers (e.g., Bajra, Jowar) emasculation is done by dipping the flowers in hot water for a certain duration (1-10 minutes) of time.

The time varies from species to species. This method Is based on the fact that gynoecia can withstand the hot temperature at which the anthers are killed. In this method an equipment is used which is placed on a simple heavy stand.

It consists of a cylindrical metallic container of 60 cm length, with one hole of 5 cm to 16 cm diameter on one end to pass over a bajra or jowar head. After inserting the panicle inside the container a cork is fitted in the hole to close it.

A 35 cm long rubber tube or belt is stretched over the side of the container, and when in use this tube is tied around the peduncle of the head. To measure the temperature, in the upper side of the container a thermometer is placed. In the field water is carried in a thermos jug.

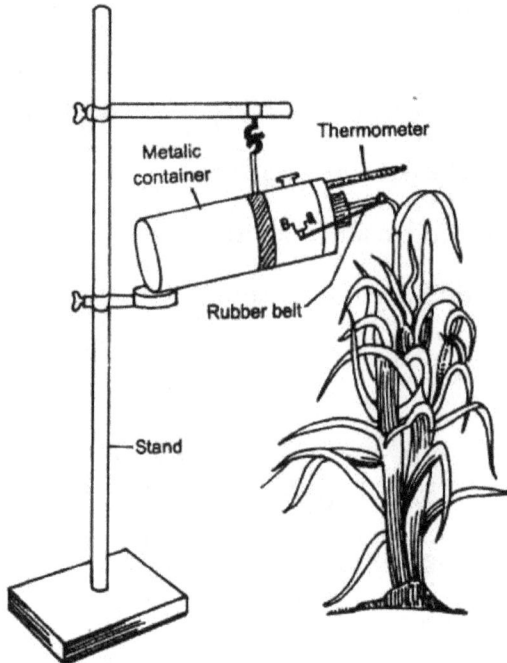

Hot water equipment for emasculation.

The panicle is inserted in the container prior to blooming for a particular duration of time. It has been observed that pollen grains of rice are killed by immersing the inflorescence for 5 to 10 minutes in the hot water maintained at 40-44° C in a thermos flask.

(c) Cold Water Treatment

Like hot water cold water also kills pollen grains without damaging the gynoecium. In rice 0-6° C temperature is maintained to kill the pollen grains. This method is less effective than hot water treatment.

(d) Alcohol Treatment Method

This method is not commonly used for emasculation because duration of treatment is an important factor since a very short duration is required failing which even the gynoecium may be damaged. Flowers or inflorescences are immersed in alcohol of a suitable concentration for a brief period. In alfa-alfa, a treatment of even 10 seconds with 57 % alcohol is sufficient to kill the pollen grains.

(e) Suction Method

It is a mechanical method and is suitable for the crops having minute flowers. In this method the amount of pressure is applied in such a way that only anthers are sucked out and other parts of the flower like gynoecium remain intact. However, in this method 10-15% self pollination takes place. It is one of the major drawbacks of this method.

(f) Male Sterility or Self-incompatibility Method

Emasculation option can be eliminated by the use of male-sterile plants, In some self-pollinated plants for example, Sorghum, Onion, Barley etc. anthers are sterile and do not produce any viable pollens! Similarly self-incompatibility may also be used to avoid emasculation.

(g) Chemical Gametocides

Certain chemicals are capable of causing male sterility, when sprayed before flowering e.g., 2, 4-D, naphthalene acetic acid (NAA), maleic-hydrazide (MA), tribenzoic acid etc. FW450 in cotton may be used for bringing about emasculation.

Bagging

It is the fourth step and completed with emasculation. The emasculated flower or inflorescence is immediately bagged to avoid pollination by any foreign pollen. The bags may be made of paper, butter paper, glassine or fine cloth. Butter paper or vegetable parchment bags are most commonly used.

The bags are tied to the base of the inflorescence or to the stalk of the flower with the help of thread, wire or pins. The bagging is done with the emasculation in bisexual plants and before the stigma receptivity and dehiscence of the anthers in unisexual plants. Both male and female flowers are bagged separately to prevent contamination in male flowers and cross-pollination in female flowers.

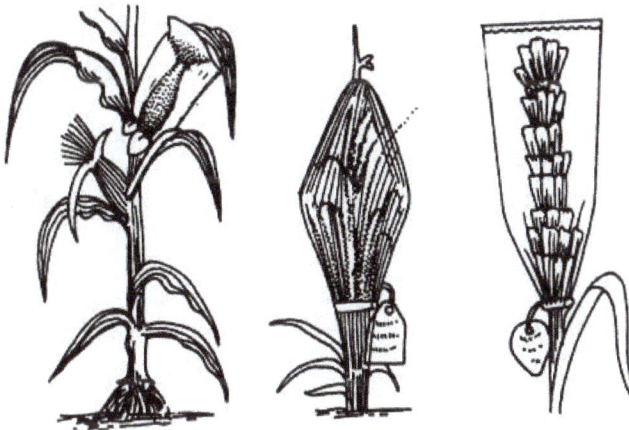

Different methods of bagging

Tagging

The emasculated flowers are tagged just after bagging. Generally circular tags of about 3 cm or rectangular tags of about 3 x 2 cm are used. The tags are attached to the base of flower or inflorescence with the help of thread.

The information on tag must be as brief as possible but complete bearing the following information:

i) Number referring to the field record.

ii) Date of emasculation.

iii) Date of crossing.

iv) Name of the female parent is written first followed by a cross sign (x) and then the male parent, e.g., C x D denotes that C is the female parent and D is the male parent.

Crossing

It is the sixth step. It can be defined as the artificial cross-pollination between the genetically unlike plants. In this method mature, fertile and viable pollens from the male parent are placed on the receptive stigma of emasculated flowers to bring about fertilization.

Pollen grains are collected in petridishs (e.g., Wheat, cotton etc.) or in paper bags {e.g., maize) and applied to the receptive stigmas with the help of a camel hair brush, piece of paper, tooth pick or forceps. In some crops (e.g., Jowar, Bajra) the inflorescences of both the parents are enclosed in the same bag.

Harvesting and Storing the F_1 Seeds

Crossed heads or pods of desirable plants are harvested and after complete drying they are threshed. Seeds are stored properly with original tags.

Raising the F_1 Generation

In the coming season, the stored seeds are sown separately to raise the F_1 generation. The plants of F_1 generation are progenies of cross seeds and therefore are hybrids.

References

- Plant-cell-chromosomes: tutorvista.com, Retrieved 31 March 2018

- 4-main-examples-of-gene-linkage-in-plants-and-animals-112760: shareyouressays.com, Retrieved 15 May 2018

- What-is-the-role-mutation-in-plant-breeding-and-evolution-2011080310003: preservearticles.com, Retrieved 21 March 2018

- Gene-interactions-allelic-and-non-allelic-cell-biology-38795: biologydiscussion.com, Retrieved 17 June 2018

- Reading-penetrance-and-expressivity: lumenlearning.com, Retrieved 19 April 2018

- Hybridization-method-of-crop-improvement-17701: biologydiscussion.com, Retrieved 29 March 2018

Chapter 5

Selection Methods and Breeding

The mode of reproduction in a plant is responsible for its genetic composition. This in turn determines the suitable breeding or selection method in the plant. The aim of this chapter is to explore the different selection and breeding methods of self-pollinated, cross-pollinated, hybrid and clonally propagated plants.

Breeding of Self-pollinated Plants

Pollination

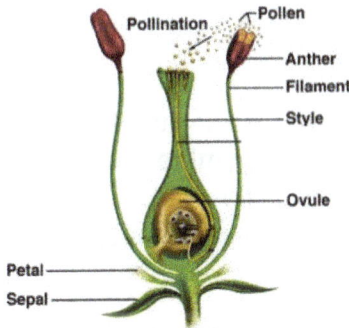

Pollination: the transfer of pollen from the anther to the sticky stigma by wind, animals/insects or water.

Self-pollination refers to a flower that has the ability to pollinate itself and other flowers produced by the same plant. It does not need another plant's pollen for fertility.

Hermaphrodites and monoecious species, which make up the majority of plant species, have the potential to self-pollinate.

A self-pollinating flower has both the male and female reproductive parts. Pollen will accumulate on the plant's male anthers and then transfer to the female stigma of the flower to complete the fertilization process.

Normally, plants that can self-pollinate feature distinctive carpels and stamens that are almost all the same length. Plants with the ability of self-pollination have the upper hand over plants that require cross pollination.

Self-pollinating plants can be pollinated by insects such as moths or bees, but they can also rely solely on the wind, airflow, or natural growth of the single flower to ensure that pollination occurs.

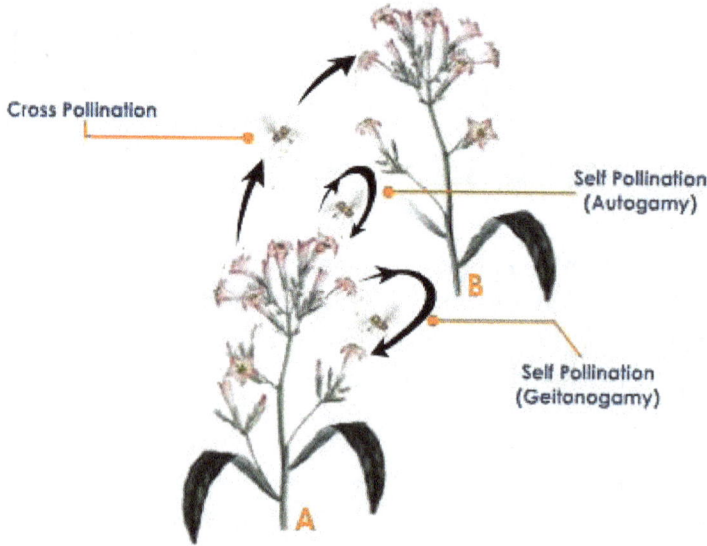

Some self-pollinating flowers can even be pollinated before they bloom by depending on the bud's natural growth. As the flower's stigma grows through the anthers and the flower begins to unfurl, it effectively transfers pollen between the flower's male and female parts.

Self-pollinated species have a genetic structure that has implication in the choice of methods for their improvement. They are naturally inbred and hence inbreeding to fix genes is one of the goals of a breeding program for self-pollinated species in which variability is generated by crossing. However, crossing does not precede some breeding methods for self-pollinated species.

The breeding methods that have proved successful with self-pollinated species are:

1. Mass selection;

2. Pure-line selection;

3. Hybridization, with the segregating generations handled by the pedigree method, the bulk method, or by the backcross method; and

4. Development of hybrid varieties.

Mass Selection

Mass selection is often described as the oldest method of breeding self-pollinated plant species. However, this by no means makes the procedure outdated. As an ancient art, farmers saved seed from desirable plants for planting the next season's crop, a practice that is still common in the agriculture of many developing countries. This method of selection is applicable to both self-and cross-pollinated species.

Key Features

The purpose of mass selection is population improvement through increasing the gene frequencies of desirable genes. Selection is based on plant phenotype and one generation per cycle is needed. Mass selection is imposed once or multiple times (recurrent mass selection). The improvement is limited to the genetic variability that existed in the original populations (i.e., new variability is not generated during the breeding process). The goal in cultivar development by mass selection is to improve the average performance of the base population.

Applications

As a modern method of plant breeding, mass selection has several applications:

1. It may be used to maintain the purity of an existing cultivar that has become contaminated, or is segregating. The off-types are simply rogued out of the population, and the rest of the material bulked. Existing cultivars become contaminated over the years by natural processes (e.g., outcrossing, mutation) or by human error (e.g., inadvertent seed mixture during harvesting or processing stages of crop production).

2. It can also be used to develop a cultivar from a base population created by hybridization, using the procedure described next.

3. It may be used to preserve the identity of an established cultivar or soon-to-be-released new cultivar. The breeder selects several hundreds (200–300) of plants (or heads) and plants them in individual rows for comparison. Rows showing significant phenotypic differences from the other rows are discarded, while the remainder is bulked as breeder seed. Prior to bulking, sample plants or heads are taken from each row and kept for future use in reproducing the original cultivar.

4. When a new crop is introduced into a new production region, the breeder may adapt it to the new region by selecting for key factors needed for successful production (e.g., maturity). This, hence, becomes a way of improving the new cultivar for the new production region.

5. Mass selection can be used to breed horizontal (durable) disease resistance into a cultivar. The breeder applies low densities of disease inoculum (to stimulate moderate disease development) so that quantitative (minor gene effects) genetic effects (instead of major gene effects) can be assessed. This way, the cultivar is race-non-specific and moderately tolerant of disease. Further, crop yield is stable and the disease resistance is durable.

6. Some breeders use mass selection as part of their breeding program to rogue out undesirable plants, thereby reducing the materials advanced and saving time and reducing costs of breeding.

Procedure

The general procedure in mass selection is to rogue out off-types or plants with undesirable traits. This is called by some researchers, negative mass selection. The specific strategies for retaining representative individuals for the population vary according to species, traits of interest, or creativity of the breeder to find ways to facilitate the breeding program. Whereas rouging out and bulking appears to be the basic strategy of mass selection, some breeders may rather select and advance a large number of plants that are desirable and uniform for the traits of interest (positive mass selection). Where applicable, single pods from each plant may be picked and bulked for planting. For cereal species, the heads may be picked and bulked.

Steps

The breeder plants the heterogeneous population in the field, and looks for off-types to remove and discard

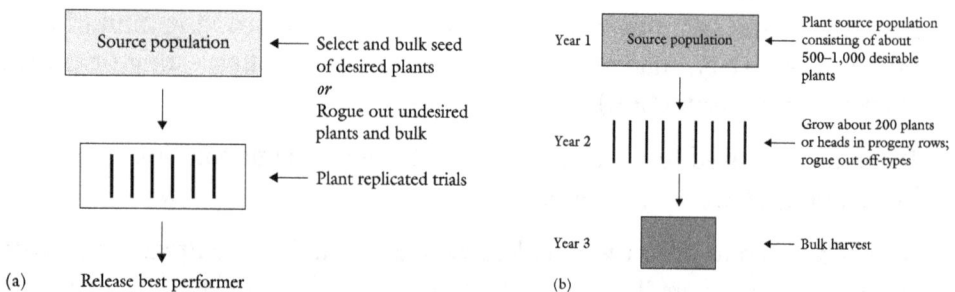

Figure: Generalized steps in breeding by mass selection for
(a) cultivar development, and (b) purification of an existing cultivar.

This way, the original genetic structure is retained as much as possible. A mechanical device (e.g., using a sieve to determine which size of grain would be advanced) may be used, or selection may be purely on visual basis according to the breeder's visual evaluation. Further, selection may be based on targeted traits (direct selection) or indirectly by selecting a trait correlated with the trait to be improved.

- Year 1 If the objective is to purify an established cultivar, seed of selected plants may be progeny-rowed to confirm the purity of the selected plants prior to bulking. This would make a cycle of mass selection have a 2-year duration instead of 1 year. The original cultivar needs to be planted alongside for comparison.

- Year 2 Evaluate composite seed in a replicated trial, using the original cultivar as a check. This test may be conducted at different locations and over several years. The seed is bulk harvested.

Genetic Issues

Contamination from outcrossing may produce heterozygotes in the population.

Unfortunately, where a dominance effect is involved in the expression of the trait, heterozygotes are indistinguishable from homozygous dominant individuals. Including heterozygotes in a naturally selfing population will provide material for future segregations to produce new off-types. Mass selection is most effective if the expression of the trait of interest is conditioned by additive gene action.

In self-pollinated populations, the persistence of inbreeding will alter population gene frequencies by reducing heterozygosity from one generation to the next. However, in cross- pollinated populations, gene frequencies are expected to remain unchanged unless the selection of plants was biased enough to change the frequency of alleles that control the trait of interest.

Mass selection is based on plant phenotype. Consequently, it is most effective if the trait of interest has high heritability. Also, cultivars developed by mass selection tend to be phenotypically uniform for qualitative (simply inherited) traits that are readily selectable in a breeding program. This uniformity not withstanding, the cultivar could retain significant variability for quantitative traits. It is helpful if the selection environment is uniform. This will ensure that genetically superior plants are distinguishable from mediocre plants. When selecting for disease resistance, the method is more effective if the pathogen is uniformly present through- out the field without "hot spots".

Some studies have shown correlated response to selection in secondary traits as a result of mass selection. Such a response may be attributed to linkage or pleiotropy.

Advantages and Disadvantages

Some of the major advantages and disadvantages of mass selection for improving self-pollinated species are given here.

Advantages

1. It is rapid, simple, and straightforward. Large populations can be handled and one generation per cycle can be used.

2. It is inexpensive to conduct.

3. The cultivar is phenotypically fairly uniform even though it is a mixture of pure lines, hence making it genetically broad-based, adaptable, and stable.

Disadvantages

1. To be most effective, the traits of interest should have high heritability.

2. Because selection is based on phenotypic values, optimal selection is achieved if it is conducted in a uniform environment.

3. Phenotypic uniformity is less than in cultivars produced by pure-line selection.

4. With dominance, heterozygotes are indistinguishable from homozygous dominant genotypes. Without progeny testing, the selected heterozygotes will segregate in the next generation.

Modifications

Mass selection may be direct or indirect. Indirect selection will have high success if two traits result from pleiotropy or if the selected trait is a component of the trait targeted for improvement. For example, researchers improve seed protein or oil by selecting on the basis of density separation of the seed.

Pure Line Selection

Pure line selection is a method in which new variety is developed by selection of single best plant progeny among traditional varieties or land races.

Pure line is a self pollinated descendent of a self pollinated plant.

Desirable types already exist in population. Those are isolated through careful testing procedures. Another term used for this method of plant breeding is individual plant selection, as large numbers of plants are selected, but those are harvested individually, their individual progenies are grown and evaluated and then best progeny is released as a pureline variety.

Genetic Basis for Pureline Selection

Pureline varieties are homozygous and homogeneous as they are genetically similar and true breeding. Such varieties possess narrow genetic base so they are more susceptible to diseases, and have poor adaptability. A pure line breeding method is normally used for self-pollinated crops, but has importance in breeding of inbred lines that are used to make hybrids in self or cross pollinated crops.

Main steps in the production of pureline variety are as follows:

• Selection of individual plants from a local / traditional variety, land races or some other mixed population

• Visual evaluation of individual plant progenies

• Yield trials

• Release of new pureline variety

Detailed Procedure

Step number	Year	Details
One	First year	Many individual plants (about 200-3000) are selected from a local variety or some other mixed population based on phenotype. Any number of plants can be selected but importantly they should be homozygous. Seeds are harvested separately.
Two	Second year	Seeds from each plant are grown separately with proper spacing. Evaluated for characters under consideration. Probably in this first season it will not be possible to judge the value of separate lines well, but surely visually poor, weak and defective progenies can be rejected. Number of progenies is drastically reduced in this step, fewer are carried to step three.
Three	Third year	Seeds of second year crop are planted in a preliminary yield trial. Standard plots for comparison are introduced, undesirable progenies are rejected.
	Four to seven years	Replicated yield trials conducted at several locations. Standard plots for comparison are introduced, undesirable progenies are rejected. Tests are done for characters under consideration.
	Eighth year	Best progeny is released as a new pureline variety. Seed is multiplied for distribution.

Examples of Pureline

Crop	Variety	Features	Selection from and at
Cowpea	Suvita-2	Striga resistant line	Selection from landrace in Burkina Faso
Wheat	Kanred	Better in winter hardiness, rust resistance, earliness in maturity	Selected at Kansas from Crimean variety, an introduction from Russia
Oat	Columbia	Matures early, grain with high test weight.	Selected at Missouri station from Fulghum variety

Advantages of Pureline Selection

Merits of pureline selection are listed below

- Easy and cheap method of crop improvement
- Rapid method, lines are usually genetically fixed and yield trials can be immediately conducted.
- Plants in such variety react in similar fashion to environmental conditions, means they are uniform in performance and at the same time in appearance too.

- Maximum possible improvement over the original variety can be achieved.

- Useful in improving low heritability traits as selection is based on progeny performance.

Disadvantages of Pureline Selection

Demerits of pureline selection are given below:

- Pure lines have poor adaptability due to narrow genetic base, just opposite to mass selected variety.

- Superior genotypes can only be isolated from the mixed population. This selection is powerless to bring changes in hereditary factors i.e. to develop new genotype.

- Mostly popular or in fact limited to self pollinated spp. only

- Time and space consuming.

- More expensive yield trials have to be conducted than in mass selection.

Applications of Pureline Selection

Applications of pureline selection are listed below:

- As pure line selection gives out uniformity in maturity, height etc. suitable for cultivars in which machines are used for the various production processes like harvesting.

- Because of the same reason as above it can be used for varieties grown for processing market, as uniformity in texture, canning qualities is important in here.

- It is practiced after hybridization within segregating populations.

- Though it can serve above purposes this method has limited practical use in the breeding of major cultivated species. However, the method is still widely used while breeding less important species that have not yet been heavily selected.

- Logically it was first step in the development of a uniform variety for the newly domesticated crops, with some amount of variability.

Hybridization

There are several methods of improvement of self-fertilized crops by hybridization. These are:

1. Pedigree method or breeding.

2. Bulk method or breeding.

3. Single seed descent method.

4. Back cross method.

5. Multiple cross method.

Pedigree Method

Individual plant progeny is selected from F_2 and subsequent generations, and their progenies are tested. During this process the record of parents as well as off-springs is kept, for which it is known as pedigree method.

The pedigree is defined as the description of the ancestors of an individual and it is generally helpful in finding out the amount of relatedness among two individuals, i.e., whether they are related by common parent in their descent ancestry or not.

Procedure

First Year

The hybridization is done among two selected parents, after emasculation one become female parent and another male parent. After seed set and maturation, the F_1 seeds are harvested separately from each plant individually. On the basis of choice of parents, the type of cross will be of two types — it will be simple cross or complex cross.

Second Year

F_1 generation seeds are space planted and selfing is allowed, each F_1 will produce more F_2 seeds. From 15-30 selected F_1 plants, the F_2 seeds are collected to get a reasonable size of F_2 population and variation.

Third Year

In F_2 generation, 2000-10000 plants are space planted, 100-500 plants are selected and their seeds are harvested separately. If the parent plants are closely related varieties then the number of selected F_3 plants would be smaller whereas in case of distantly related varieties the number of F_3 progenies will be of relatively larger numbers.

Fourth Year

In F_3 generation also, the individual plant progenies are space planted. Each progeny should have about 30 or more plants. Individual plants with desirable characteristics are selected; disease and lodging susceptible progenies to be eliminated, and also the progenies with undesirable characters are rejected even from the selected plants.

During this selection if the number of superior progenies is very small then the whole cross programme may be rejected.

Fifth Year

The selection procedure is same as previous year, only if two or more progenies coming from the same F_3 progeny are similar and comparable, then only one may be saved and others may be rejected. The emphasis is given on the selection of desirable plants from superior progenies.

Sixth Year

Individual plant progenies of F_5 generation are planted according to recommended commercial seed rate. Three or more rows for each progeny will help in comparison among progenies. Many progenies may have become reasonably homozygous genotype and may be harvested in bulk. If the progenies show variation then the individual plants are selected. The number of selected progenies should be reasonable so that preliminary yield trial with 25-100 progenies can be done.

Seventh Year

Individual plant progenies of F_6 generation are planted in multi-row plots and evaluated visually. Progenies harvested in bulk since they become homozygous. The segregating progenies may be discarded and the preliminary yield trial may be done for the progenies which are reasonably homozygous and have enough seeds.

Eighth Year

Preliminary yield trial with three or more replications is conducted to identify few superior lines. The progenies are evaluated for plant height, lodging, disease resistance, flowering time, maturity time, etc. Quality test is done to serve as an additional basis for selection.

Ninth to Tenth or Thirteenth Year

The superior lines are tested in replicated yield trials at several locations. The above mentioned criteria are evaluated for these lines. The line which is superior to the best commercial variety may be released as new variety.

Eleventh or Fourteenth Year

The selected strain should get multiplied to release as a new variety. Breeder has the responsibility to supply the seeds to the state seeds corporation for production and marketing of the seeds.

Schematic Representation

First Year		Selected plants of parental variety are hybridized to get the F1 hybrid seeds.
	P₁ P₂	
Second Year		The hybrid seeds (10-30) are planted and the seeds of F2 harvesed n bulk.
	F₁	
Third Year		2000-1000 seeds are space planted. 100-500 superior plants are selected and seeds are harvested from those as pedigree
	F₂	
Fourth Year		Individual plant progeries are planted in raw and space planted. Superior plants are selected
	F₃	
Fifth Year		The same procedure as in fourth year.
Sixth Year		The individual selected plant progenies are in multi-row plots.
	F₅	
Seventh Year		The same procedure as in previous year Enough seeds to be collected from superior plants.
	F₆	
Eight Year		The Preliminary yield tiral and quality test has to be done.
	F₇	
Ninth to Thirteenth Year		Co-ordinated yield trials to be done. Disease and quality tests have to be done.
	F₈ – F₁₂	
Fourteenth Year		Multiplication of seeds for distribution and release as a new variety.
	F₁₃	

Merits

1. This method is most useful as transgressive segregation for yield and other quantitative characters may be recovered in addition to improvement of specific characters.

2. This method is well suited for improvement of characters, which can be easily identified and simply inherited.

3. Through the maintenance of pedigree record the breeder may be able to obtain the information about inheritance of characters.

4. Plants or progenies with weaker and visible defects are eliminated at an early stage in the breeding programme.

5. This method gives maximum importance on the breeder to use his/her skill and judgement about the selection of plants and progenies.

6. This method takes less time than bulk method to release a new variety.

Demerits

1. The success of the method is mainly dependent on the skill of the breeder.

2. To keep the individual pedigree record is laborious and time consuming, it may be the limiting factor for large breeding programme.

3. Selection of large number of progenies in every generation is also laborious and time consuming.

4. In F_2 and F_3, the selection for yield is not effective. If sufficient number of progenies is not retained, valuable genotypes may be lost in early segregating generations.

Achievements

Pedigree method is useful in selection of new superior recombinant types from a hybridization programme. This method is suitable for improving specific characteristics, such as disease resistance, plant height, maturity time, etc. as well as yield and quality characters.

Many improved varieties have been developed through pedigree method in many crops like wheat, rice, barley, pulses, oil seeds, cotton, tobacco, jowar, vegetables, etc.

Wheat

K 65 (tall variety) ← C 591 x NP 773.

K 68 (good quality grain) ← NP 773 x K 13.

WL 711 (dwarf, high yield) ← (S308 x Chris) x Kalyan Sona.

Malviya 12 (good grain) ← NP 876 x Cno 66.

Rice

'Jaya' and 'Padma' (short duration, finer grain) ←Taichung Native 1 x T 141.

Cotton

'Laxmi' (fibre quality, early maturing, resistant to leaf blight) Gadag 1 x CC 2 (Cambodia Coimbatore 2).

Tomato

Pusa early dwarf (more yield) ← Meerut x Red cloud.

Bulk Method

It is a method which can handle segregating generations, in which F_2 and subsequent generations are harvested in bulk to grow the next generation. At the end of bulking period, individual plant selection and evaluation is carried out in the similar fashion as in the pedigree method.

Procedure for Bulk Method

Bulk method of plant breeding completes in following main steps.

Step	Details
Hybridization	Crossing among selected parents
F_1 generation	F_1 seeds (minimum 20) planted. Bulk harvesting is carried out.
F_2 to F_6 generation	F_2 to F_6 are planted, harvested in bulk. Number of plants should be as large as possible. Generally artificial selection is not carried out.
F_7 generation	Generally 30 to 50 thousand F_6 seeds are space planted, selection is carried out based on phenotype and 1000 to 5000 selected ones are harvested separately.
F_8 generation	Individual plant progenies are grown, inferior progenies eliminated. Harvested in bulk.
F_9 generation	Preliminary yield trials with standard varieties as check. Selection is based on yield.
F_{10} to F_{13} generation	Multi location yield trials are conducted using standard varieties as check. Evaluation of performance of lines is done.
F_{14} generation	Seed multiplication for distribution.

Advantages of Bulk Method

Here are the merits of bulk method.

- Simple and inexpensive.

- Little record keeping.

- Easy to handle populations as harvest in bulk.

- Natural selection is effective, possible to carry out artificial selection if required.

- We can wait for opportunity to carry out selection. Selection for some environment dependent characteristics like disease resistance, lodging resistance can be carried out when environment is favorable for disease epidemic and severe lodging. Till that year the bulk can be maintained easily.

Disadvantages of Bulk Method

Demerits of bulk method are given here:

- Requires more time duration to develop new variety.

- Pedigree record is not maintained, so we can not trace back progeny to the parent plant.

- Not suitable method for greenhouses.

- Large number of progenies has to be selected at the end of bulking period.

- Natural selection may also work against desirable traits.

- Not much scope for skills.

- Bulk method is particularly suited to small grain crop plants.

- Maximum productivity is established in F_2 generation of method.

- No recombination occurrence among superior lines.

Applications of Bulk Method

- Homozygous lines can be isolated in less time, automatic increase in homozygosity up to F_6 or F_8 generation after this individual plant selection can be carried out.

- Natural selection may improve considered character (yield), with minimum expenses.

- Suitable for crops which are generally planted at high planting densities, e.g. small grain crops.

Single Seed Descent Method

This method was suggested by Coulden for advancing segregating generation of self-pollinated crops. A breeding procedure used with segregating populations of self-pollinated species in which plants are advanced by single seeds from one generation to the next is referred to as single seed descent method.

Single Seed Descent

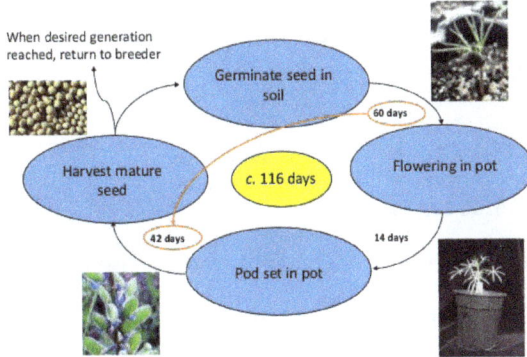

The procedure is as follows:

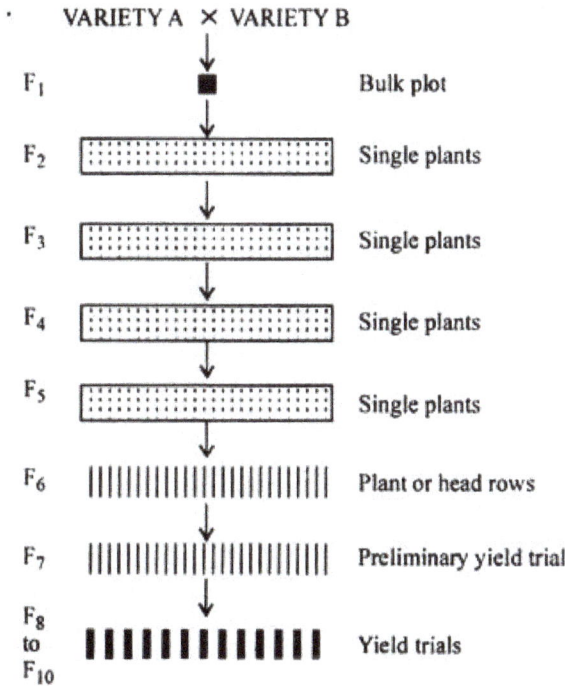

Figure: Single seed-descent method of selection.

Advantages

- It is an easy and rapid way to attain homozygosity.

- Small spaces are required in early generations to grow the selections.

- Natural selection has no effect.

- The duration of the breeding program can be reduced by several years by using single seed descent.

- Every plant originates from a different F2 plant, resulting in greater genetic diversity in each generation.

- It is suited to environments that do not represent those in which the ultimate cultivar will be commercially produced.

Disadvantages

- Natural selection has no effect.

- Plants are selected based on individual phenotype not progeny performance.

- Inability of seed to germinate or plant to set seed may prohibit every F2 plant from being represented in the subsequent population.

- The number of plants in the F2 is equal to the number of plants in the F4. Selecting a single seed per plant runs the risks of losing desirable genes. The assumption is that the single seed represents the genetic base of each F2. This may not be true.

Back Cross Method

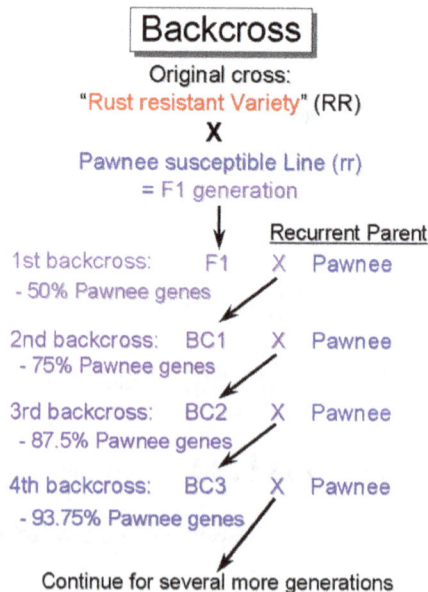

Backcross

Original cross:
"Rust resistant Variety" (RR)

X

Pawnee susceptible Line (rr)
= F1 generation

Recurrent Parent

| 1st backcross: | F1 | X | Pawnee |

- 50% Pawnee genes

| 2nd backcross: | BC1 | X | Pawnee |

- 75% Pawnee genes

| 3rd backcross: | BC2 | X | Pawnee |

- 87.5% Pawnee genes

| 4th backcross: | BC3 | X | Pawnee |

- 93.75% Pawnee genes

Continue for several more generations

The backcross procedure is most often used to move a transgene from a good tissue culture variety that was used in transformation to an elite experimental line or variety. It turns out that for many crops, once the transgene is in the crop species crossing is more efficient than transformation procedures. Backcrossing is more efficient than transforming the elite line because most transformation protocols are optimized for a specific (often poorly adapted and lower yielding) laboratory line. Many elite lines (which are high yielding) are not amenable for transformation. Hence genetic engineers transform their lab line and breeders backcross the transgene from the lab line into the elite line.

The backcross method is a form of recurrent hybridization (repeated crossing to a single variety) where a superior characteristic may be added to an otherwise desirable variety. In this method the breeder has considerable control of the genetic variation in the segregating population in which selections are to be made. The backcross method has been used extensively for transferring qualitative characters (characters with clear phenotypes that are easy to identify in cross progeny) such as disease resistance. It is effective in both self and cross pollinated crop species. To better understand the applications of backcrossing, the gene for leaf rust resistance in wheat will be used as an example. Figure shows the visible symptoms of leaf rust in susceptible wheat. The left picture leaves are resistant, they have a small amount of rust. The plant on the right is covered with rust, it is susceptible.

Figure: Two examples of wheat, the one on the left is resistant to rust,
the one on the right is susceptible to rust.

Figure: illustrates how the backcross procedure can be used to move leaf rust resistance (RR, Rr) from one variety to a susceptible variety (rr).

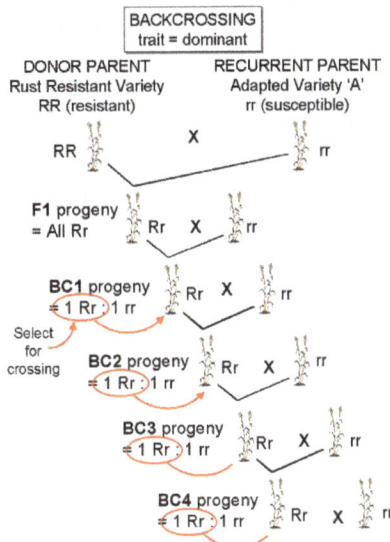

Figure: Backcross breeding with a dominant trait.

The actual procedure for back crossing is almost self-explanatory. In back crossing you have a donor parent (has a gene of interest) and a recurrent parent (an elite line that could be made better by adding the gene of interest). The donor parent is crossed to the recurrent parent. The progeny of this cross is then crossed to the recurrent parent (it is 'crossed back' to the recurrent parent, hence the term back cross). The progeny of this cross is selected for the trait of interest and then crossed back to the recurrent parent. This process is repeated for as many back crosses as are needed to create a line that is the recurrent parent with the gene of interest from the donor parent. The goal of backcrossing is to obtain a line as identical as possible to the recurrent parent with the addition of the gene of interest that has been added through breeding.

In the end, you want to keep only the individuals homozygous for the resistance gene. To obtain them, self Rr plants from BC_4. The resulting offspring will be 1RR : 2Rr : 1rr. Progeny testing would be needed to identify RR from Rr plants. Progeny testing is where the genotype of a parent plant is determined by genotypes of the line's progeny. In the case of an RR plant, the progeny will all be RR (no segregation for the gene/trait). However in the case of an Rr plant, the progeny will segregate 1/4 RR : 1/2 Rr : 1/4 rr. Therefore, the progeny of RR plants will be uniformly resistant to leaf rust, while the progeny of Rr plants will segregate for resistance and susceptibility.

In contrast, if the genes for rust resistance had been recessive (i.e., ss = resistant) rather than dominant, then the introduced resistant gene is only carried in the heterozygote and would not be detected throughout the backcross program. After each backcross, one would have to self the heterozygote (Ss) in order to produce resistant plants (ss) in the progeny. These resistant plants (ss) are then backcrossed to the recurrent parent (SS).

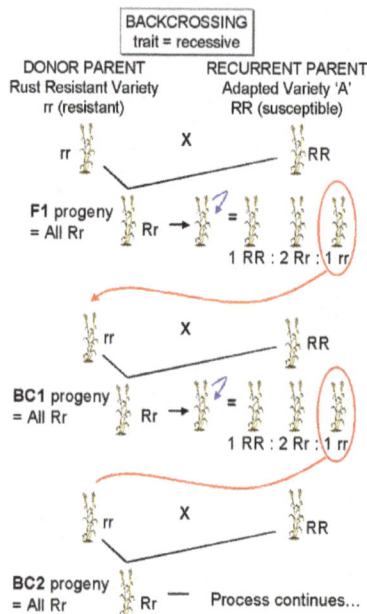

Figure: Backcross breeding with a recessive trait.

When working with recessive traits, such as this example, Allard suggests advancing the 1st backcross to the F_2 generation followed by selection for the desirable character from the donor parent (ss) and the general features of the recurrent parent. The 2nd and 3rd backcrosses are then made in succession after which the inbreeding with selection phase for ss is repeated. This is followed by the 4th and 5th backcrosses in succession. The BC_5F_2 that are resistant (SS) are crossed to recurrent parent (SS) for the BC_6F_1 which is Ss. The BC_7F_1 is selfed to get in the BC_6F_2 : 1/2 SS (susceptible) : 1/2 Ss (susceptible) : 1/2 ss (resistant) backcross with intense selection for both the desired character (ss) and the recurrent parent plant phenotype. You have successfully transfered the gene.

Applications

Generally the backcross method is used for transferring disease resistance character to a good and well adapted variety. Other quantitative characters can also be transferred through back-cross method in both cross- and self-pollinated crop.

1. Inter-varietal Transfer of Simply Inherited Characters

The characters like disease resistance, seed color, plant height, etc., which are controlled by one or two major genes are most suited for transfer through back-cross method from one variety to another of the same species. The successful transfer depends on the minimum linkage between desirable and undesirable trait.

2. Inter-varietal Transfer of Quantitative Characters

Grain characters like seed size and shape, earliness, plant height-all can be transferred from one variety to another with the criteria of high heritability.

3. Interspecific Transfer of Simply Inherited Characters

Mainly the character like disease resistance can be transferred from related species to cultivated species. This specially requires the relatedness between the species where the chromosomes can pair during meiosis. The transfer will be unsuccessful if the genetic environment of recurrent parent is not suitable for functioning of the gene of desirable character from donor parent.

4. Transfer of Cytoplasm

In case of transfer of male sterile character from one parent to another requires back-cross method. The variety or species from which the cytoplasm is to be transferred is used as female parent. The recurrent parent should be the male parent. After 6-8 back-crosses, the cytoplasm will be of the donor parent with the genotype of recurrent parent.

5. Transgressive Segregation

Modification of backcross method will produce transgressive segregants. Few (1 to 3) back-crosses with F_1 allow much heterozygosity to appear, or two or more recurrent parents may be used in back-cross programme to accumulate genes from them into the back-cross progeny. Such kind of modification would produce new variety which won't be like the recurrent parent.

6. Production of Isogenic Lines

Backcross method is useful for production of isogenic lines, i.e., the lines of a crop, which are identical in their genotype except for one gene. These isogenic lines are useful for studying the effect of individual genes.

Merits

1. This method does not change the genotype of the popular established variety, only it helps a single desirable character to be transferred in the existing variety.

2. As the recurrent parent is an established variety, so it is not necessary to test the yield performance which ultimately saves five years-time, as well as expenses.

3. In case of short duration plant, 2-3 generations can be raised within a year as the selection is based on inheritance of a particular character, not on performance. So this method drastically reduces the time period required to develop a new variety.

4. Backcross method requires smaller populations than the pedigree method.

5. Defects of an established variety can be removed using this method, may be only by introducing a single character.

6. Interspecific gene transfer only can be done through this method.

7. Transgressive segregation can be obtained in case of quantitative characters by modified backcross method.

8. This method is very much useful for cytoplasmic gene transfer to the recurrent parent, i.e., new variety will bear the cytoplasm of donor parent and genotype of recurrent parent.

Demerits

1. The new variety is not superior in performance than the existing variety except the introduction of single character.

2. During transfer of such kind of single desirable gene, sometimes some undesirable genes may get transferred.

3. For introduction of more than one gene controlled character, multiple crossing programme is necessary which is often difficult, time taking and costly.

4. Once a recurrent parent taken in a backcross method may get replaced by another superior variety of high yielding ability.

Achievements

Backcross method is very much useful for transferring of simply heritable character like disease resistance to well adapted popular local variety.

In case of wheat, 'Kalyan Sona' is the popular established variety to which the leaf rust resistance character has been transferred from diverse sources like Robin, KI, Bluebird, Tobari, etc. using back-cross method.

In case of Bajra, Tift 23A, a male sterile line which was susceptible to downy mildew has made resistant through back-cross method.

For interspecific transfer of genes, back-cross method is widely used for crop improvement. Cultivated sugarcane (Saccharum officincirum) is susceptible to pests and disease, crossed with S. spontanium which is resistant. This brings resistance but with undesirable characters like more fibre, low sugar, thin stem, etc. By back-crossing with noble cane these undesired characters are removed.

In cotton, the hybridization between Gossypium hirsutum and G. arboreum yielded highly sterile F hybrid, few tetraploid seeds were obtained. These plants were then back- crossed with G. hirsutum and two varieties have been selected from the back-cross progeny which are now being widely cultivated in Gujarat.

Except the characters of disease resistance other characters can also be transferred by the back-cross method. Such as in cotton, the ginning quality of fibre has been improved by back-cross method from the local variety. By using BD8 (wilt resistant, high spinning value but low ginning out turns) as recurrent parent and Goghari A26 (high ginning out turn) as non-recurrent parent, new variety Vijay was developed.

Multiple Cross Method

A cross-involving more than one inbred line is referred to as multiple cross. It is also known composite cross and is used to combine monogenetic characters from different sources into a single genotype. In this method, several pure lines are crossed together. The selected pure lines are first combined into crosses as A × B, C × D, E × F, G × H and so on.

The F_1 plants are mated together as $(A \times B) \times (C \times D)$ and $(E \times F) \times (G \times H)$. Finally, the F_1 plants of double crosses are crossed with each other to produce hybrids $[(A \times B) \times (C \times D)] \times [(E \times F) \times (G \times H)]$. Further breeding in these hybrids is carried out according to either pedigree or bulk method.

$$A \times B \quad C \times D \quad E \times F \quad G \times H$$
$$AB \times CD \qquad EF \times GH$$
$$ABCD \quad \times \quad EFGH$$
$$ABCDEFGH$$

Merits

1. In self-pollinated crops this method is used when three or four monogenic characters scattered in three or four different varieties are to be combined into one.

2. These crosses generally have wider adaptation.

Demerits

1. These crosses are generally less productive.

2. This method has limited utility except in high risk areas where severe disease damage occurs regularly from a highly specialized disease pathogen.

Hybrid Varieties

The development of hybrid varieties differs from hybridization in that no attempt is made to produce a pure-breeding population; only the F_1 hybrid plants are sought. The F_1 hybrid of crosses between different genotypes is often much more vigorous than its parents. This hybrid vigour, or heterosis, can be manifested in many ways, including increased rate of growth, greater uniformity, earlier flowering, and increased yield, the last being of greatest importance in agriculture.

By far the greatest development of hybrid varieties has been in corn (maize), primarily because its male flowers (tassels) and female flowers (incipient ears) are separate and easy to handle, thus proving economical for the production of hybrid seed. The production of hand-produced F_1 hybrid seed of other plants, including ornamental flowers, has been economical only because greenhouse growers and home gardeners have been willing to pay high prices for hybrid seed.

Recently, however, a built-in cellular system of pollination control has made hybrid varieties possible in a wide range of plants, including many that are self-pollinating, such as sorghums. This system, called cytoplasmic male sterility, or cytosterility, prevents normal maturation or function of the male sex organs (stamens) and results in defective pollen or none at all. It obviates the need for removing the stamens either by hand or by machine. Cytosterility depends on the interaction between male sterile genes ($R + r$) and factors found in the cytoplasm of the female sex cell. The genes are derived from each parent in the normal Mendelian fashion, but the cytoplasm (and its factors) is provided by the egg only; therefore, the inheritance of cytosterility is determined by the female parent. All plants with fertile cytoplasm produce viable pollen, as do plants with sterile cytoplasm but at least one R gene; plants with sterile cytoplasm and two r genes are male sterile (produce defective pollen).

The production of F_1 hybrid seed between two strains is accomplished by inter planting a sterile version of one strain (say A) in an isolated field with a fertile version of another strain (B). Since strain A produces no viable pollen, it will be pollinated by strain B, and all seeds produced on strain A plants must therefore be F_1 hybrids between the strains. The F_1 hybrid seeds are then planted to produce the commercial crop. Much of the breeder's work in this process is in developing the pure-breeding sterile and fertile strains to begin the hybrid seed production.

Breeding of Cross-pollinated Plants

The transfer of pollen grains form the anther of one flower to the stigma of another flower belonging to the same species or closely allied species is called cross-pollination or Allogamy. When the pollination takes place between two flowers of the same species, it is called xenogamy, while if it occurs between two closely related species, it is called hybridism.

Characteristics of Cross-pollination

Cross pollination by insect

Cross-pollination usually occurs in plants having unisexual flower. It may also occur in flowers showing male sterile lines e.g. maize, Solanum. Flowers showing different maturation times for stamens and carpels may also show cross-pollination, e.g. sunflower, Magnolia. It may be in bisexual flowers with differential stamens and carpels, commonly known as heteromorphism e.g., Oxalis. Some flowers, where a barrier is created between stamen and carpel also shows cross-pollination e.g. Iris. The flowers showing different genetic makeup are showing the cross-pollination.

Adaptations for Cross-pollination

a.   Dicliny or Unisexuality: In unisexual flowers, self-pollination is impossible, so cross-pollination is observed e.g. gourd.

b.   Self-sterility : In certain flowers, the pollen grains are sterile in nature due to male sterility and hence they cannot fertilize the egg of the same flower, so they depend on the pollen grains of another flower coming via cross-pollination, e.g. Solanum; maize.

c.   Dichogamy: The androecium and gynoecium of a bisexual flower do not mature at the sametime, so self-pollination can never take place and they are of two types: Protandry (The anther matures earlier than the stigma, e.g. sunflower) and Protogyny (The stigma matures earlier to anther, e.g. Polyalthia).

d.   Herkogamy: In this case, self-pollination is impossible, because some of the floral parts act as a physical barrier between anther and stigma and thus cross-pollination is favored. In Iris flower, the anthers are extrorse and they are concealed by overlapping or overarching of the petaloid style, which conceals the anther, making self-pollination impossible.

e.   Heteromorphism: The flowers of one single species may vary on the basis of the forms of stamens and carpels and accordingly, the flowers may be dimorphic or trimorphic in nature.

f.   Homomorphicself incompatibility: It is a rare type of self incompatibility, in which stamens and pistil do not vary in size, but even than, the pollen grains are rejected by the stigma of the same flower. It is found in Oenothera, Aster.

Methods of Cross-Pollination

There are various types of cross-pollination and the agents are discussed below:

1. Anemophily

When pollination is brought upon by wind, it is called anemophily and the flowers are called anemophilous, e.g. paddy, wheat, maize and grasses.

Adaptations of wind-pollinated flowers: These types of flowers show the following characteristics:-

1. The flowers are small and not easily seen.

2. The petals are not colored and they are not scented i.e., devoid of osmophores.

3. They are without nectaries.

4. The flowers are aggregated on a long peduncle above the vegetative parts, which makes the process of wind-pollination easy.

5. The sepals and petals are small and not easily seen and sometimes, undifferentiated to form perianth.

6. The accessory whorls do not cover the sexual reproductive organs.

7. The stamens are provided with long filaments with versatile anthers, which are easily cut off by air current.

8. The pollen grains are small, granular, light weight, dry and shaped in huge quantity.

9. The style is also long and that helps in the protrusion of the stigma from the flower.

10. The stigma is large, feathery and branched, which helps in easy trapping of the pollen grains.

2. Hydrophily

When the cross-pollination in a flower takes place with the help of water, it is called hydrophilous and the phenomenon is termed as hydrophily. The water pollinated plants are of two types:

i.   Hypohydrogamous: The pollination taking place in completely submerge condition under water e.g. Ceratophyllum.

ii. Epihydrogamous: The pollination taking place along the surface of the wateje.g. Vallisneria and Hydrilla.

Adaptations for Water-pollinated Flowers

The hydrophilous flowers show the following characteristics:

1. The flowers are small, inconspicuous, light in weight, helping in floatation.

2. The flowers are not showy, without colored petals, without any fragrance.

3. The floral parts are covered with waxy substance or cutin, which prevent them from getting damaged by water.

4. The accessory whorls, calyx and corolla are small, so the essential floral whorls or androecium and gynoecium are always exposed in water current.

5. The dehiscence of anther is rapid and so the pollen grains are scattered in wider areas in a short time.

6. The pollen grains are small, light in weight, impervious to water and hence carried by water to long distances.

7. The female flowers usually have a short coiled stalk, that reach the water surface by uncoiling.

8. The stigma is provided with bristles, which can easily trap the pollen grains floating in water.

9. The coiled stalk of female flower may recoil again after pollination.

Method of Pollination

In hypohydrogamous flowers like Ceratophyllum, the flowers never comes above the water surface, the male flowers have superior position, which drop the pollen-grains (impervious to water) on to the stigma of the female flower remaining below. The epihydrogamous flowers are always at the water surface, so pollination is always brought about on the water surface, where pollen grains are carried by the water-current from the male flower, up to the stigma. Some flowers, like that of Vallisneria, the flowers usually remain submerged. The small.male flowers on maturity get detached from the spadix and float on the water surface on a boat-like spathy bract. The coiled stalk of the female flower uncoils and the female flowers come to the water surface, their stigmas come in contact with the anthers of the male flower and pollination takes place. After pollination, the stalk of the female flowers recoil again and the flowers again go down underwater.

3. Zoophily

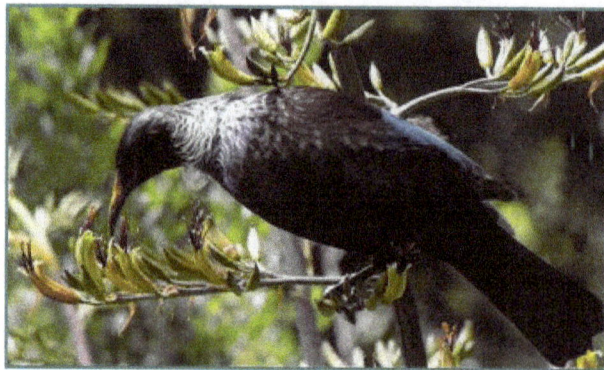

Cross pollination by bird

When the pollination is brought about by animals, the pollination is called zoophily and the flowers are called zoophilous e.g. Calotropis. The zoophilous flowers in general shows the following characteristics:

1. The flowers are brightly coloured, showy, sweet scented, i.e., with osmophores.

2. They may contain the nectaries.

3. The pollen grain are edible and palatable.

4. The exines of the pollen grains are rough and sticky, which easily get adhered to the body surface of animals.

5. The stigma surface is usually sticky to trap the pollen grains after coming into its contact.

The cross-pollination brought about by insects is called entomophily and the flowers are called entomophilous e.g. Salvia, sunflower, Calotropis. The pollination affected by birds is called ornithophily and the flowers are called ornithophilous. e.g. BignoniaJSterilitzia. When the pollination is brought about by snails and slugs, it is called malacophily and the flowers are called malacophilous e.g. members of Aroideae. When the process of cross-pollination brought about by bats is called chiropteriphily and the flowers are called chiropteriphilous. e.g. silk cotton, Bauhinia.

The top four breeding methods used for cross-pollinated crops. The methods are:

1. Mass Pedigree Method.

2. Inbreeding.

3. Recurrent Selection.

4. Synthetic Varieties.

Mass Pedigree Method

In this method of breeding, the best individuals with desired characters are selected on the basis of phenotypic performance in a source population. Open pollinated seeds of the selected individual plants are divided into two halves. "Second year replicated progeny row trial is conducted using one set of half seeds from each plant.

On the basis of the progeny performance, the best parental individuals are identified. The remnant half seeds from the superior parental plants are mixed and grown in isolation for random mating during the third year.

This method of breeding is equivalent to ear-to-row selection in context of maize originally proposed by C.G. Hopkins at the Illinois Agricultural Experiment Station in 1896 to improve protein and oil content of maize. This method has been named as mass pedigree method by S.S. Rajan in India. This very method is called line breeding when selection is based on progeny tests and a group of progeny lines is composited.

Inbreeding

The mating of individuals more closely related than individuals mating at random is known as inbreeding. The lines produced by continued inbreeding are known as inbred lines. Self-fertilization is the most intense form of inbreeding.

In plant breeding nearly homozygous lines are produced by continued self-fertilization accompanied by selection for five to six generations. This can be used as the method of breeding only in those crops, which do not show any loss of vigor due to inbreeding, like cucurbits.

The three important uses of inbreeding in cross-pollinated crops are as follows:

1. To attain uniformity in plant characters.

2. To improve yield etc. by individual plant selection as in cucurbits in which there is no inbreeding depression.

3. To develop suitable inbred lines in production of hybrids and synthetics.

Synthetic Variety

The term 'synthetic variety' has come to be used to designate a variety that is maintained from open pollinated seed following its synthesis by hybridization in all combinations among a number of selected genotypes, which have been tested for combining ability.

The components of a synthetic variety could be inbred (usually), clones, mass selected populations or various other materials. The component units are maintained so that the synthetic may be reconstituted at regular intervals.

The inbreeds to be used as component lines are chosen on the basis of combining ability tests. The components inbred are crossed in all possible combinations. This inter-crossed seed is called as Syn 0.

Equal quantity of seed from all crosses is composited and the mixture is allowed open-pollination in isolation and seed is harvested. This becomes Syn 1 generation. In absence of reconstitution of a synthetic at regular intervals, the population becomes an open-pollinated variety.

The testing for combing ability is the decisive criterion for a synthetic variety by which it can be distinguished from a conventional variety of a cross-pollinating species, which originates in a continuous selection of individuals and subsequent progeny tests. The greater variability caused by crossing several components with high general combining ability makes the synthetic varieties more adaptable compared to conventional varieties.

Similar to hybrids, the yield of a synthetic variety generally also decreases after the Syn 2, until an equilibrium is reached which, in partially self-fertile species, depends on selfing rate and inbreeding (minimum depression), but also on the number of components used in the Syn 0.

The performance of Syn 1 can be estimated by the formula:

$$\hat{F}_2 = \overline{F}_1 - \left(\frac{\overline{F}_1 - \overline{P}}{n}\right)8$$

The yield to be expected, usually increases with the number of components until an optimum is reached. The question regarding the most favorable number of genotypes the Syn 0 should be composed of cannot be clearly answered, because the evidence from research and practice is too divergent.

The yield can be increased by:

1. An increase of the mean performance of the F1 combinations.

2. An increase of the mean performance of the parents (inbred lines, clones).

3. An optimum combination of the components.

Therefore, it is obvious that I_0 plants should already be tested for their combining ability and plants or lines should be used as components, the inbreeding depression of which is not as strong as in the I_5. In order to maintain performance in subsequent generations, mass selections have been found to be sufficient in maize.

The performance of synthetics can be improved by one further breeding cycle. It consists of selection of genotypes from the synthetic variety, their testing by a dialled, and combination of genotypes with the highest combining ability for a new synthesis (= recurrent selection).

Composite Varieties

Concept of composite varieties (in maize) originated in India. Composite varieties are generally derived from the varietal crosses in advanced generation. These are usually developed from open-pollinated varieties or other heterozygous populations or germ plasm which have originally not been subjected to inbreeding or have not been elaborately tested for their combining ability.

Usually, they involve open pollinated varieties, synthetics, double crosses, etc., selected for yield performance, maturity, resistance to diseases and pests. These composites often show a high order heterosis in F_1's when widely diversed populations are crossed. Advanced generations of such heterotic crosses often show stabilized yields. General combining ability and additive gene effects play predominant role in exploitation of these populations.

The details of the steps involved in development of composite variety are as follows:

1. Screening of diverse germ plasm by evaluation at multi-locations/years to identify the sources having adaptability, desirable agronomic attributes and resistance to major diseases and tolerance/resistance to serious insects.

2. Making of all possible crosses among selected superior genotypes or top crossing with a varietal complex of screened base varieties.

3. Conducting multi-location test with the F_1 and F_2 generations of varietal crosses and selection of F_2's showing desirable agronomic features along with least decline in F_2.

4. Evaluation of selected F_2 populations and identification of the best one as practically composites are constituted by compositing seeds of various populations and allowing the mixture to stabilize under open pollination along with some mild selection in isolation. The constituent entries may not be maintained for reconstituting the composite. Composite may serve as a base population for developing inbred lines.

Hybrids

The term hybrid variety is used to designate F_1 populations that are used to commercial planting. The F_1's are obtained by crossing genetically unlike parents. The pioneering work on hybrid maize was done by G.H. Shull (single crosses) and D.F. Jones (double crosses). Other innovative researchers in this area have been E.M. East, H.K. Hayes, F.D. Richey and M.T. Jenkins and others.

Types of Hybrids

(i) Single Cross

A single cross is a hybrid progeny from a cross between two unrelated inbred.

= A x B

(ii) Three-way Cross

A three-way cross is the hybrid progeny from a cross between a single cross and an inbred.

(A x B) x C

(iii) Double Cross

A double cross is the hybrid progeny from a cross between two single crosses.

$$(A \times B) \times (C \times D)$$

(iv) Modified Single Cross

A modified single cross is the hybrid progeny from a three- way cross, which utilizes the progeny from two related inbred as the seed parent and an unrelated inbred as the pollen parent.

$$(A \times A') \times B$$

(v) Double Modified Single Cross

A double modified single cross is the hybrid progeny from two single crosses, each developed by crossing two related inbred.

$$= (A \times A') \times (B \times B')$$

(vi) Modified Three-way Hybrid

A modified three-way hybrid is the progeny of a single cross as female parent and another single cross between two related inbred.

$$= (A \times B) \times (C \times C')$$

(vii) Top cross Hybrid

This is inbred x variety hybrid. Following top cross hybrids may be formed:

   a.   Inbred line x variety
   b.   Inbred line x experimental hybrid
   c.   Inbred line x synthetic variety
   d.   Inbred line x family

(viii) Double Top Cross Hybrid

A double top cross hybrid is the progeny of a single cross and a variety. Such hybrids have been produced on commercial scale in India and China.

Methods of Inbred Line Development

Standard Selfing Method

Self-Pollination of individual plants within single plant progenies grown is the most common procedure used to develop inbred lines. This breeding procedure has two important problems.

(a) Vigor of the lines is decreased with inbreeding because of loss of favorable dominant allelles and any heterozygous loci that have over-dominant effects. Many lines are so poor in seed yield, pollen production, etc., that they cannot be used in a programme to produce single cross hybrid seed.

(b) Effective selection within the row for the plants that have desired agronomic traits becomes minimal in generations beyond S_3. Therefore, some breeders use only two or three generations of self-pollination with subsequent reproduction by sib-mating within progenies.

Special Techniques

These techniques involve the doubling of haploids derived from either maternal or paternal gametes. Somaclonal variations from inbred lines offer another opportunity. However, due to low frequency of haploids and doubling of haploids to the diploid state, these methods are still not an important component of most breeding programmes.

Besides, these two methods, there are three other methods where along with developing inbred lines, there are opportunities to improve them simultaneously.

These are as follows:

Pedigree Method

In this method, a pair of elite lines that complement one another are crossed to produce the F_2 generation and pedigree selection is practiced by sampling the F_2 population. F_2 populations of the single crosses are the most frequently used source populations for line development.

Backcrossing

This is a modification of pedigree method. Modifications of backcross method have also been suggested for example, in convergent improvement by Richey, there is parallel improvement of two inbred lines by the reciprocal addition of dominant favourable genes present in one line and lacking in the other line.

In this method, two inbred A and B are crossed. The F_1 is backcrossed with A followed by selection of desirable traits of B and F_1 is also backcrossed with B where selection for desirable traits of A is made. After about three backcrosses and selection, selfing is done to fix the selected genes. This method is useful for improving such characters as vigour, resistance to diseases, pests and lodging.

Gamete Selection

This scheme devised by L.J. Stadler in 1944, is based on the premise that if superior zygotes occur with a frequency of p^2, superior gametes would occur with a frequency of p. The procedure involves crossing an elite line with a random sample of pollen of plants from a source population.

The resulting F_1 plants and the elite line are testcrossed to a common tester and F_1 plants are also selfed. Testcross progenies are evaluated in a replicated trial. The test crosses of F_1 plants that exceed the elite line by tester are presumed to have obtained superior gametes from the source population. Superior gametes are recovered as F_2 self's.

Selection and selfing are continued till desirable homozygosity/uniformity is attained. Though gamete selection is not used as extensively as pedigree and backcross methods, it does have some intrinsic features, and consequently, is included in some breeding programmes.

Testing of Inbred

The general combining ability of the inbred is tested by making all possible crosses [n (n – 1)/2] in a dialed fashion or else top cross test is carried out. Specific combining ability is also estimated.

Time of Testing

One procedure is to test the inbred for hybrid performance/combining ability in about the fifth generation of selfing when the number of selected lines is greatly reduced. This breeding system assumes favorable relationships of plant and other traits of inbred lines (traits as selection criteria during inbreeding) with combining ability for grain yield.

A second system of inbred development is based on an evaluation for hybrid perfor-mance in the early generation of self- pollination, e.g., testcrosses of the S_0 plants or S_1 lines. Genotypes that are identified for above average performance in these tests are continued in the selfing and selection nursery.

This procedure has been called as 'early testing' originally proposed by M.T. Jenkins in 1935. The assumption is that the combining ability of a line is determined early in its development and will change relatively little in subsequent generations of inbreeding and selection.

By early testing, the breeder is in a position to discard some lines that are inadequate in hybrid performance and wasteful expenditure on these lines is avoided. However, probably most breeders use a method that is intermediate between these two systems.

In this approach, first hybrid evaluation is of S_2 or S_3 lines.

Combination of Inbred in Hybrids and Prediction of Double Cross Performance:

   Single Crosses: n (n − 1) / 2

   Three-way crosses: n (n − 1) (n − 2) / 2

   Double crosses: n (n − 1) (n − 2) (n − 3) / 8

   where, n = Number of inbred

If there are 4 inbred, A, B, C and D, the performance of the double cross (AxB) (CxD) is predicted as follows:

   i.   Based on mean performance of all possible six single crosses:

      (A x B) + (A x C) + (A x D) + (B x C) + (B x D) + (C x D)/6

   ii. Based on mean performance of the four non-parental single crosses:

      (A x C) + (A x D) + (B x C) + (B x D)/4

   iii. Based on mean performance of four lines over a series of single crosses:

      (A x E) + (A x F) + (A x G) + (A x H) + (B x E) + (B x F) + (B x G) + (B x H)+.......... +(D x H)/n

   iv. Based on mean performance of top-crosses of the four inbreeds:

      (A x variety) + (B x variety) + (C x variety) + D x variety/4

Of the above four methods, method (ii) is found to give more accurate results.

Genetical Basis of Heterosis

The phenomenon of hybrid vigour, expressed particularly in the first generation (F_1) following the crossing of cultivars or inbred lines, has been known for more than a hundred years. The term heterosis, coined by G.H. Shull in 1909 suggests a mechanism based on heterozygosity and therefore, is not fixable in the homozygous state. Two hypotheses have been put forward to explain heterosis.

They are as follows:

(i) Over-dominance Theory

This was proposed by G.H. Shull and E.M. East independently in 1908. According to this hypothesis, hybridity/heterozygosity is superior to either homozygote and this state of heterozygosity has a stimulating effect upon the physiological activities of the organism leading to superiority of Aa over AA or aa.

(ii) Dominance Theory

This hypothesis was proposed by C.B. Davenport in 1908, A.B. Bruce in 1910 and F. Keeble and C. Pellew in 1910. According to this hypothesis, each dominant allele contributes equally to heterosis and the recessive alleles contribute nothing. It is also assumed that the dominance is complete.

If all the dominant alleles are concentrated in one parent, and the counterpart recessive alleles in another parent, the F_1 will be equal to the parent having all the dominant alleles.

For example:

```
    AABBCCDD                      aabbccdd
       P₁              ×              P₂
                       ↓
       F₁        AaBbCcDd
```

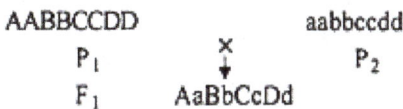

In the above cross, if each dominant allele, contributes 1 unit and the recessive allele, 0 unit, then the P_1 will have a value of 4 and P_2, a value of 0. F_1 will have a value of 4.

However, there are situations, where F_1 is superior over the better parent.

These cases under dominance theory can be explained assuming that dominance and recessive alleles are distributed in both the parents as given below:

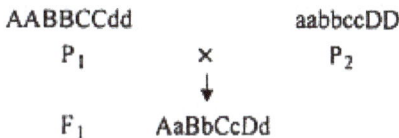

```
    AABBCCdd                      aabbccDD
       P₁              ×              P₂
                       ↓
       F₁        AaBbCcDd
```

In this cross, the phenotypic value of P_1 is 3, that of P_2 is 1 and F_1 has a score of 4 which is superior to the better parent. Under this model it should be possible to derive a pure line from the F_1 which should be equal to F_1 in the performance and thus this heterosis will be fixable.

However, it is generally agreed that heterosis is not fixable in the homozygous state. This would be the case if hybrid vigor were due to true over-dominance or due to tight linkage in the repulsion phase at some incompletely dominant loci.

Much evidences suggest that apparent over-dominance is, in fact, due to non-allelic interaction and linkage disequilibrium and that heterosis is mainly a result of the bringing together of unidirectionally dominant alleles distributed between the parental line. Under this, heterozygosity is not an essential prerequisite for high performance, uniformity and stability of performance.

Recurrent Selection

Recurrent selection is a method of breeding designed to concentrate favorable genes scattered among a number of individuals by selecting in each generation among progeny produced by matings inter-se of the selected individuals (or their selfed progeny) of the previous generation.

Based on the ways in which plants with desirable characters are identified, recurrent selection has been divided into four types.

These types are:

1. Simple recurrent selection or recurrent selection for phenotype.

2. Recurrent selection for general combining ability.

3. Recurrent selection for specific combining ability.

4. Reciprocal recurrent selection.

In simple recurrent selection a number of plants are self-pollinated in a source population in first year. At maturity superior plants based on phenotypic performance are selected. In second year, seeds produced by self-fertilization of the selected plants are planted and crossed in all possible combinations and the produce is bulked.

This completes original selection cycle. Since selection is based on the phenotype of the plant, it is useful only for characters with high heritability. In those cases, where it is possible to identify the desired selections before flowering as in case of cauliflower, cabbage, etc., inter-crosses of selections may be made in the first year of each cycle and the second year may be eliminated from each cycle.

Thus, strictly speaking, selfing is not an integral component of simple recurrent selection, rather it is done only to prevent crossing from the inferior pollen grains before the plants reach to selection stage.

In recurrent selection for general combining ability, a three year cycle is involved. In first year a number of plants are self-pollinated and crossed to a broad based heterozygous tester stock to identify the S_o plants with good general combining ability. In second year, the crosses are evaluated to identify those that are superior. Self's of first year are kept in reserve.

In third year, the reserve selfed seeds are grown out, inter-crossed in all combinations, and a composite of inter-crossed seeds is used to establish an improved population for further selection. This procedure developed as a direct outgrowth of studies of early testing first proposed by M.T. Jenkins in 1935.

Recurrent selection for specific combining ability was proposed by F.H. Hull in 1945. This method of selection is same as that of recurrent selection for general combining ability except that the tester selected is a narrow base an inbred line. The recurrent selection for general and specific combining ability is equivalent to half sib progeny test.

Reciprocal recurrent selection proposed by R.E. Comstock, H.F. Robinson and P.H. Harvey in 1949 aims at simultaneous improvement of two heterozygous and heterogenous populations (designated as A and B).

A serves as tester for B and B serves as tester for A. This method is as effective as recurrent selection for gca when additive gene action predominates, and is as effective as recurrent selection for sea when non-additive effects are of major importance.

The steps are as follows:

I. Season:

Selected plants of population A are self-pollinated and crossed to plants of population B. Likewise plants are selected and self-pollinated in B and outcrossed to plants of population A.

II. Season:

Test cross progenies of both the populations are evaluated in replicated trial. Superior progenies are identified on the basis of performance in this trial.

III. Season:

Selfed seed from plants with superior test cross progenies are grown population wise separately and inter-crossed to reconstitute two populations, which will be now called as A' and B'. This completes one cycle and additional cycles may be initiated.

Synthetic Varieties

Synthetic varieties are open pollinated populations developed through random mating of selected genotypes. Synthetic varieties were first suggested by Hayes and Garber and defined by Lonnquist as open pollinated varieties (OPVs) derived from the intercrossing of selfed plants or lines known to possess high general combining ability (GCA), and subsequently maintained by routine mass selection procedures from isolated plantings.

Ultimately the commercial cultivars may be open-pollinated populations (synthetics or composites) and hybrids. The choice of cultivar depends upon resources, stage of breeding program, infrastructure and manpower for seed production and socioeconomic factors. Hybrid cultivars have the advantage of higher yield potential and uniformity. They are preferred over open-pollinated populations subject to higher level of heterosis. However, seed production of hybrids is costlier and a bit tedious and complicated.

Synthetics and the specialized populations derived from them-known as synthetic cultivars (also commonly referred to as synthetic varieties, which are considered completely equivalent to synthetic cultivars here) are common products of plant breeding activities in a wide array of cross-pollinated species. Various definitions have been applied to these populations and some plant breeders have considered them to be equivalent, although this can lead to confusion. Following Lonnquist, a synthetic is an open-pollinated population maintained in isolated plantings that is derived from the random mating of selfed plants or lines or other genotypes (parents) produced from mass selection. As such, a synthetic is simply the bulked seed resulting from one or more cycles of population improvement that involve artificial selection.

Lonnquist defined a synthetic as an open – pollinated population formed by intercrossing of selfed plants or lines and subsequently maintained by mass selection. The term synthetic variety has come to be used to designate a variety that is maintained from open-pollinated seed following its synthesis by hybridization in all possible combinations among a number of selected genotypes which have been subjected to combining ability test. The components of a synthetic variety could be inbreds (usually) or mass selected populations in context of maize. The components are maintained so that the synthetic variety could be reconstituted as and when required.

Developing synthetic varieties through the use of full-sib and half-sib families or clones as parents is a commonly used breeding method in alfalfa. Synthetic varieties have become increasingly favored in alfalfa and other forage crops, mainly because it is cheaper than the development and use of hybrid varieties. Developing synthetic varieties also helps to minimize productivity loss with advancing generations of seed increase.

A synthetic variety is developed through intercrossing of several genotypes of known superior combining ability. Genotypes selected for synthetic variety development are those that are known to give superior hybrid performance when crossed in all combinations. Thus, properly selected male and female parents from diverse origin that can increase the possibility of heterosis when crossed are essential to successfully develop synthetic varieties. Therefore, before selecting clones for developing synthetic varieties, breeders should test the clonal progenies from polycross in yield trials.

In maize first inbreds are developed. The inbreds to be used as component lines are selected on the basis of combining ability for which component inbreds are crossed in all possible combinations. The intercrossed seed is called as So seed. Equal quantity of seed from all crosses is mixed and the mixture is allowed open-pollination in isolation and the seed is harvested. The harvested seed represents S1 generation. In absence of reconstitution of a synthetic variety at regular intervals, the synthetic variety becomes an open-pollinated variety.

The overall performance of the synthetics depend mainly on the number of parental components, their general combining ability, their specific combining ability and on the total amount of heterosis. Synthetics can be used by either farmers for commercial production or breeders as source populations from which to select new lines. To utilize part of the heterosis in faba bean, synthetic cultivars were recommended for spring beans in Europe; the advantages of these cultivars are not only their partial use of heterosis, but also the possible increase in yield stability.

To utilize and increase the involvements of synthetic varieties in crop improvements full information of this issue is very important. But still there is no organized information on the synthetic population, their source, selection methods, developments and their application in agriculture. So this review is objected on providing of detailed materials on the above mentioned issues.

Synthetic Populations

Improved varieties are synthetic cultivars, usually obtained through three or four generations of open pollinated reproduction of polycross seeds of selected parents. This is a great way to create diverse new landraces and open-pollinated lines. Synthetic populations do better than regular open-pollinated types but are not as high yielding as F1 varieties.

Synthetic populations and the composite crosses described below are generally ways to generate diversity from which you can select, either by bulking or recurrent mass selection. Synthetics may also be bred as a goal in itself and the first initial crosses grown together in a mixture that is remade each year (or after several years). To create a synthetic population:

1. Start with two or more variable landraces or heirloom varieties.

2. Make all possible hybrid combinations between all plants. For example, if you started with Brandywine and Green Zebra tomatoes, you would make a hybrid using Brandywine as the mother and Green Zebra as the father (pollen source) and vice versa.

3. Pool all of the seeds of the hybrids together and plant them out.

4. Allow these plants to naturally and randomly homogenize and mix.

5. Over the next three to five generations, select out the best plants.

In obtaining synthetic varieties in mixed species, the reasoning is broader than in allogamous varieties. In the latter, the general and specific combination capacity matter. In mixed species, maintenance of a synthetic variety involves crosses and natural self-pollination. The ideal is thus to select superior inbred lines derived from parents that also have good combination capacity.

Synthetic variety breeding is most effective and intensive method to improve perennial forage crops like alfalfa through polycross. Classical breeding studies require long time to select individual clones for synthetic variety production. Synthetic varieties are populations that are generally created by intermating a set of proven inbred lines.

Synthetics can achieve higher yield levels than older open-pollinated varieties that have received little systematic breeding for yield. They can also be propagated for many generations with little loss of yield. Several studies have shown that synthetics currently produce lower yields that are probably not acceptable to most farmers. The open-pollinated synthetic populations developed for breeding purposes generally produce lower yields, often one-third less than conventional hybrids. Some of these populations do have a high protein content in their grain (9-13%) relative to conventional hybrids (7-9%) so that protein and essential amino acid yields may be similar on a per acre basis. Such open-pollinated corn should have a higher feeding value and price as organic protein from other sources is expensive.

A synthetic variety is developed by intercrossing a number of genotypes known for superior combining ability with high genetic distance. Therefore, synthetic variety is made up of genotypes previously tested for their ability to produce a superior progeny when crossed in all combinations in agreement with Ferreira et al. (1995) who emphasised that heterosis and the combining ability of parents depend directly on the genetic diversity between them and the chance of finding promising combinations is better when more divergent material is used.

Populations can be a result of crosses among a set of homozygous inbred lines (synthetic varieties), an open-pollinated variety, or a mixture of varieties and races (composites). General theories, however, make no distinction about the origin of the population unless it does not fill some of the basic requirements. Synthetic variety constitutes a poly morphical and stable population. Hence synthetic varieties are a high adaptation to environment variations. In other words synthetic varieties provide stable yield in the fluctuating environment.

Sowing seed of synthetic varieties is a common practice in forage species such as alfalfa (Medicago sativa L.) and orchargrass (Dactylis glomerata L.). Such varieties from selection and crossing of lines are phenotypically uniform but different genotypes. These lines to cross freely year after year to produce seeds, heterozygous and heterogeneous populations originate. The use of artificial seed allows multiplication of outstanding genotypes and genetically uniform, since this method does not require that annually crosspollination is carried out to produce plants (McKersie and Brown, 1996). Having the advantage over other method of stable yield and less cost of production of new varieties synthetic population is very unique to be used by breeder in crop improvement.

Sources of Synthetic Populations

Synthetic variety is the same to hybrid once but development of synthetic varieties requires the development of inbred lines, their testing for general combining ability, making their all possible cross combinations, predicting the F2 performance constituting a number of experimental synthetics, testing their yield levels in yield trials over locations, and finally releasing those which excel the standard checks. In India, one of the experimental synthetic populations "Syn 65" has been released for cultivation in the name of cv. "Sangam." In lotni brown sarson (B. campestris), Pusa Kalyani has been developed and released as cultivar from the IARI research station Kanpur (UP) utilizing this breeding approach.

Examining the relationships among progeny tests (correlations, heritability, components of additive and genetic variances) is of great importance in determining which of the tests is the most suitable (effective) for the purposes of breeding and developing synthetic varieties.

Maize (*Zea mays* L.) synthetic populations are low-cost and stable varieties, obtained by cross pollination of a group of inbred lines. They are a viable alternative for situations where the use of hybrid seed and related inputs are too expensive. Although synthetic populations are generally less productive than heterotic hybrids, their main advantage is that the heterosis does not diminish significantly in F_2. Besides the inbred lines, maize synthetics can be obtained from hybrids or local populations. Obtaining maize synthetics from local populations aims at enriching the gene pool with a large number of valuable genes derived from local population characteristic to some agricultural areas.

These positive aspects come to support the approach promoted by CIMMYT in order to obtain highly productive synthetic populations of maize, which is of great importance mainly in places where the use of hybrid seed is too expensive (especially in developing countries). From that year on, synthetic maize populations have acquired a special importance as objectives of research in the field.

Breeders want to improve synthetic populations of maize for using them in obtaining superior inbred lines necessary for hybridization programmers. The value of any maize population depends on its potential per se and on its combining ability in crossings. The per se value of synthetic populations of maize has been studied for many traits: productivity, earliness, resistance to falling, resistance to Sesamia nonagrioides Lef. and Ostrinia nubilalis.

Knowledge about genetic variability in species is important for optimal use of genetic resources in plant breeding programs. The use of molecular markers especially AFLP (Amplified fragment length polymorphism) markers help to select genetic dissimilarity potential parents for production of synthetics. Whatever the type of material will be used in the development of synthetic population its combining ability is the crucial point to be considered.

Development of Synthetic Varieties

The first objective in the development of synthetic varieties is to increase the gene frequency for specific attributes (Hallauer and Eberhart. The basic concept in the development of synthetic varieties is exploitation of heterosis or hybrid vigor, such as varieties are constituted from general combining ability inbreds. However, heterosis is partially utilized by synthetic varieties because some level of inbreeding takes place to open pollination in later generations. Synthetic exploit more of additive gene action whereas hybrid exploit more non additive (over dominance and epistatic) gene action.

A synthetic variety is a variety produced by crossing in all combinations a number of in-bred lines (with high GCA that combine well with each other) and a synthetic variety is maintained by open pollination in isolation. In maize, development of synthetics includes:

- Evaluation of lines on the basis of general combining ability.

- These selected lines are intercrossed in all possible combinations.

- Equal amount of seed from these crosses is composited to constitute a synthetic.

- A synthetic variety is developed by intercrossing a number of genotypes known for superior combining ability with high genetic distance. Therefore, synthetic variety is made up of genotypes previously tested for their ability to produce a superior progeny when crossed in all combinations. Though the end goal of development of synthetic varieties is a plant with a general com-bining ability, following of some stapes and having full knowledge of all this becomes valuable issue. So breeders should ready before starting of the meth-od of improvement.

Selection Methods for Synthetic Variety

After the development of the synthetic population different methods employed to se-lect the best performing individuals. Mass selection and phenotypic selection are the selection methods for synthetic variety development. Molecular markers have been widely used for genetic diversity studies and marker assisted selection for synthetic variety development.

Selecting inbred lines based on their GCA in defined crosses is used to develop synthet-ics and improve yield when selection is directly on yield (a trait of very low heritability) has a limited effect. This suggests that while the inheritance of combining ability is a quantitative trait governed by many genes, each has a larger individual effect than would genes that contribute directly to yield. Lines with high GCA presumably have a larger proportion of favorable yield genes which differ from other lines measured with respect to their specific favorable gene complex.

Measurements of GCA can be obtained from multiple top crosses made between a line and various elite inbreds, and these crosses are phenotyped for yield in multiple loca-tions and years. Although specific combining ability (SCA) is more important from the standpoint of obtaining maximum yields in hybrid crosses, GCA is highly important in developing high yielding synthetics. Line performance in top crosses has been shown to be relatively constant after the S1 generation and consequently little or nothing would be gained by additional selfing where the production of synthetic varieties is the goal.

Open pollinated varieties developed by modern plant breeding were historically used as commercial cultivars by farmers, and replaced the older landraces that had been selected directly by the farmers. Later, the best OPVs were selected as source populations for further plant improvement and development of synthetics and modern hybrids. In the 1920s, nearly one thousand cultivars available in the United States Corn Belt were selfed in an attempt to develop useful inbred lines. These were often intercrossed to create synthetics. One of the most popular, Iowa Stiff Stalk Synthetic (BSSS), was developed by GF.

Sprague in the early 1930's by intermating 16 inbreds (Hallauer and Miranda, 1988), and is considered an excellent source population for the selection of inbred lines with high combining ability with other elite inbred lines.

THAD reported that the rate of change in inbreeding in synthetic varieties per generation, is always greater than 1/2, for diploid, equilibrium is reached in one step past SYN 1, i. e., in the SYN 2, involved from 2 to 64 parents. In order to evaluate a synthetic variety we need to predict the yield performance of SYN 2.

Practicing prior recurrent selection in the population should increase the occurrence of more homozygous mother plants with good performance in crosses and negligible inbreeding depression. This should be possible from the wide genetic variation detected. Such a fact will favor obtaining superior synthetic varieties. The amount of seeds of a synthetic cultivar is multiplied by successive generations of random mating without selection; these generations are called Syn-2, Syn-3, and so on.

Applications of Synthetic Varieties

Breeding through synthetic variety is most effective and intensive method to improve perennial forage crops like alfalfa through polycross. Classical breeding studies require long time to select individual clones for synthetic variety production. The use of molecular markers especially AFLP (Amplified fragment length polymorphism) markers help to select genetic dissimilarity potential parents for production of synthetics.

The prime synthetics have confirmed to be a precious resource of genetic variability for disease resistance. Synthetic wheat also showed resistance to leaf blotch, glume blotch, crown rot, yellow leaf spot, leaf blight, powdery mildew, karnal bunt. The vast majority of fruit species are propagated by vegetative means because of the presence of selfin compatibility and breeding cycles very long. The use of synthetic seed facilitates its spread. However, the most useful artificial seed would be in the conservation of germplasm of these species. In countries with large varied conditions like Ethiopia synthetic varieties can play a greater role to overcome the influence global warming by increasing the genetic diversity of the crop. In addition to this it can also stabilize the production and productivity as it can resist many diseases and insects.

Synthetic Genomics

Synthetic genomics has been defined as "the engineering of biological components and systems that do not exist in nature and the re-engineering of existing biological elements; it is determined on the intentional design of artificial biological systems, rather than on the understanding of natural biology". Synthetic biology aims to design and model novel biomolecular components, networks and pathways. These are then applied to rewire and reprogram organisms to provide solutions for various challenges.

One of the goals of synthetic genomics is the preparation of viable minimal genomes which will function as platforms for the biochemical production of chemicals with economic relevance. The production of biofuels, pharmaceuticals and the bioremediation of environmental pollution are expected to constitute the first commercial applications of this new technique. Synthetic biology has enabled the construction of a gene that encodes the same amino acid sequence as the plant enzyme but that is optimized for expression in the engineered microorganism of choice. This method has provided massively parallel throughput which has made it possible to identify and track genetic variation among the various strains, providing insights into why some strains are better than others (BIO, 2013). As the world is shifting toward this science, developing countries should relate themselves for better competence in crop improvement and feed their people who are in hunger of food and nutrition.

Breeding of Hybrid Cultivars

A cultivar is a clone or seed strain selected for a particular trait or traits. The value of hybrid cultivars was discovered through experiments with corn in the 1920/30s. Corn is a cross-pollinated crop. But if a corn plant is forced to self-pollinate by covering the silks (=stigmas) from external sources of wind-blown pollen and then pollinating the silks with pollen from the tassels of the same plant seed can be produced. When grown the next generation the progeny do not grow as vigorously as their parent and over many generations of enforced self-pollinating they become even weaker. This is referred to as inbreeding depression. This can be so severe that the weakest lines fail to produce seed but some inbred lines have adequate vigor to be maintained. It was discovered that if inbred lines were hybridized, in certain cases the progeny were more vigorous (larger, higher yielding) than the starting material from which they have been derived. This is known as either hybrid vigor or heterosis. This discovery eventually led to the development of the hybrid corn industry and subsequently other crops.

Whereas other cultivars breed true and their grain can be used as seed for another generation, this is not true of hybrid varieties. Cross-pollination is of course necessary among the plants (in e.g. a hybrid corn field) for grain to be produced. But if that grain is used as seed for next year's crop it will not breed true, the crop will exhibit tremendous variability in all characteristics (height, maturity etc.). In the case of canola hybrids, this would also extend to quality characteristics such as glucosinolate content, erucic acid content and any other oil quality characters of that hybrid and make the seed from that crop unacceptable. Thus farmers must obtain seed of a hybrid (which is produced under controlled conditions by the seed company) each year. Many vegetables are also sold as hybrids such as tomatoes. Saving some seed for planting next year will reveal the hidden variability of a hybrid. As in other crops, new hybrid cultivars (based on new combinations of inbred lines) are being brought to market all the time.

Top Cross Method

A top cross hybrid is defined as a hybrid that results from cross pollinating a single-cross hybrid or an inbred line with an open-pollinated variety. It is considered to be the quickest means of developing hybrids and it produces excess seeds. In addition, the cost of producing top cross hybrids is comparatively cheaper than conventional hybrid seeds. Since the cost of production is cheaper, the price of seeds is also cheaper; hence, small holder resource poor farmers can afford to buy them.

The decline in grain yield due to recycling top cross hybrid seed is half of that of classical hybrids; nonetheless, they balance well with respect to grain yield. The end result of planting "recycled" or second generation seed was insignificant for OPVs (5%), rigorous for hybrids (> 30% yield loss) and intermediate for top cross hybrids (16% yield loss). Given that many farmers in Ghana recycle seeds set aside from their harvest, the use of non-conventional top cross hybrids could be one approach of enhancing yield levels.

In this method widely used in maize, many inbred lines are chosen and back crossed with the original variety there are two types of top crosses viz, single top cross and double top cross.

In single top cross method, two different inbred lines are crossed with each other and the F, crossed with the original variety.

Individuals with desired traits are selected. In double top cross method F hybrids of two different lines are crossed and the product crossed with one of the parents.

Single Cross Method

A hybrid plant results from a cross of two genetically different plants. The two parents of a single-cross hybrid, which is also known as a F1 hybrid, are inbreds. Each seed produced from crossing two inbreds has an array (collection) of alleles from each parent. Those two arrays will be different if the inbreds are genetically different, but each seed contains the same female array and the same male array. Thus, all plants of the same single-cross hybrid are genetically identical. At every locus where the two inbred parents possess different alleles, the single-cross hybrid is heterozygous.

Plants of a single-cross hybrid are more vigorous than the parental inbred plants. In Figures 2a and 2b, the single-cross hybrid plant and ear are shown with the plants and ears of the parental inbreds. Clearly, the hybrid plant is taller and the hybrid ear is larger. The increase in vigor of a hybrid over its two parents is known as hybrid vigor.

Figure: Corn Plants: Inbred B73 (left), Inbred Mo17 (middle), Single cross B73 x Mo17 (right)

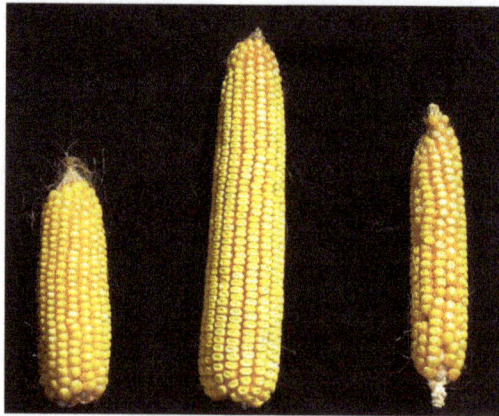

Figure: Corn Ears: Inbred B73 (left), Inbred Mo17(right), Single cross B73 x Mo17 (middle)

Breeders often measure the degree of hybrid vigor of a trait with the following formula:

$$\frac{100(Hyb - MP)}{MP}$$

where, Hyb = the value of the trait in the hybrid and

MP = the average (mid-parent) value of the trait in the two parents

For example, in figure the height of the single-cross hybrid is 3.0 m (this equals Hyb), the average height of the inbreds is 2.0 m (this equals MP), and the value of hybrid vigor is 50%. Hybrid vigor calculated in this way is called mid-parent hybrid vigor. Another type is high-parent hybrid vigor. This is the superiority, expressed as a percentage, of the hybrid over the parent with the better or higher value of the trait. Corn breeders will be successful in increasing hybrid performance if the hybrid vigor of a new hybrid compared to an older hybrid is increased and the two sets of parents have equal performance and/or if hybrid vigor is unchanged but the mid-parent value of the parents of the newer hybrid is superior to that of the parents of the older hybrid.

The genetic basis of hybrid vigor is not completely understood. However, experience has shown that a hybrid produced by crossing two inbreds that are closely related usually will exhibit less hybrid vigor than a hybrid produced by crossing inbreds that are more distantly related.

If a single-cross hybrid is allowed to open-pollinate (pollen is despersed freely), each of the plants grown from the resulting seed will be genetically unique. To understand why this is so, first consider a single locus. All plants of a single-cross hybrid are genetically identical, so at a single heterozygous locus any cross- or self-pollination occuring with open-pollination can be represented as:

$$A_1A_2 \times A_1A_2$$

One-half of the egg cells produced by each plant carries the A_1 allele and one-half carry the A_2 allele. The same is true of the pollen cells. The egg and pollen cells then combine at random during pollination.

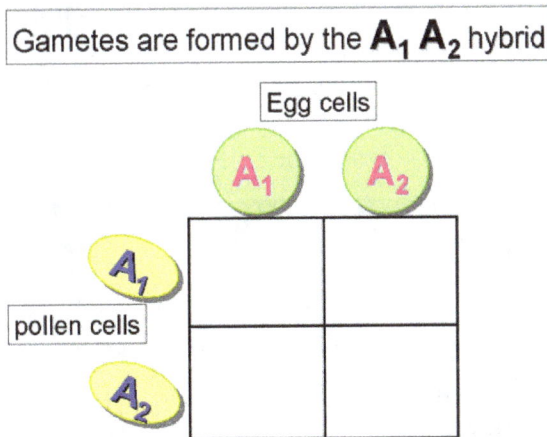

Gametes are formed by the $A_1 A_2$ hybrid

Use the diagram to predict the kinds of offspring produced from this cross.

The A_1A_2 and A_2A_1 genotypes are functionally identical, so three unique genotypes can be produced at a single heterozygous locus when open-pollination occurs.

With two heterozygous loci, A and B, the situation can be illustrated as follows:

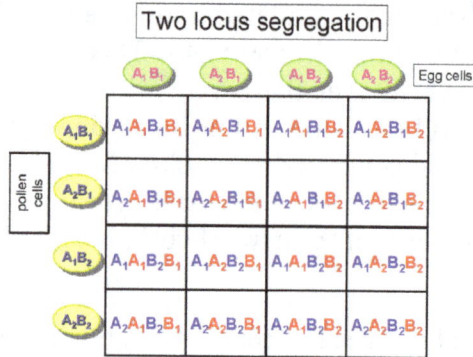

In the two-locus case, nine unique genotypes are produced. This occurrence of multiple genotypes among progeny arising from the self- or cross-pollination of parents that all have the same heterozygous genotype at one or more loci is known as genetic segregation.

This segregation occurs at hundreds or even thousands of gene loci. The number of unique genotypes resulting from genetic segregation at n loci is given by 3^n. Thus, if $n=1$ (*i.e.*, one locus), then the number of unique genotypes is three, and if $n=2$ the number of unique genotypes is nine. But, if $n=20$, the number of unique genotypes balloons to 3,486,784,401 ($=3^{20}$) Any commercial single-cross hybrid of corn is likely heterozygous at many more than 20 loci. That is why open-pollination of such a single cross results in progeny that are all genetically unique.

The progeny produced from self-pollination of a F1 single-cross hybrid are known as F2 plants. On average, F2 plants will have vigor that is approximately half-way between the single-cross parental plants and the average of the two inbred grandparents; that is, half of the hybrid vigor is lost. This is illustrated in figure. The F2 ears on the bottom row vary in size, but on average are larger than the ears from their inbred grandparents and smaller than the ear from their single-cross parent. That is why farmers have an incentive to purchase new single-cross hybrid seed each year.

Figure: Top row of corn ears: Inbred B73 (left), Single cross B73 x Mo17 (middle), Inbred Mol7 (right)
Bottom row of corn ears: From F2 plants derived from B73 x Mo17

When single-cross hybrid seed is commercially produced, one inbred is the male parent and the other the female parent. Either the female parent must be male-sterile (pollen is not produced or is not functional) or the tassel on each female plant must be removed (this is called detasseling) prior to any pollen production. In either case, all the seed produced on the female parent will be single-cross hybrid seed.

Developing an inbred from a single-cross hybrid requires approximately seven generations of repeated self-pollinations. Each year in the United States, commercial seed companies produce hundreds of new inbreds and test in field trials many thousands of new single-cross hybrids obtained by crossing these inbreds. Compared to existing commercial hybrids, the vast majority of these new hybrids will be poorer or no better in performance. Only the hybrids that have superior performance in these trials are produced in mass quantities and sold as commercial hybrids to farmers.

Figure: A single-cross hybrid production field with female inbred
parent detasseled) and male inbred parent

Considerable time and inputs are required to develop, select, and produce single-cross hybrids. Achieving a high level of cost efficiency of these processes typically requires large-scale operations.

Production of Hybrid Seed

Hybrid is produced by crossing between two genetically dissimilar parents. Pollen from male parent (Pollen parent) will pollinate, fertilize and set seeds in female (seed parent) to produce F1 hybrid seeds. For production of a hybrid CROSSING between two parents is important, the crossing process will results in heterosis. In self pollinated cross it is difficult to cross but in cross pollinated crops it is easier.

In nature to create genetic variability and for its wider adaptation in different environmental conditions, flowering plants has adopted many mechanisms for cross pollination. Cross-pollination results in genetic heterogeneity and show wider adaptations. Flowering plants have evolved a number of devises to encourage cross-pollination. Those mechanisms are:

1. Dicliny: Flowers are unisexual. In monoecious plants male and female flowers are borne on the same plant e.g., cucurbits, maize, castor and coconut. In dioecious plants male flowers are borne on different plants e.g., papaya, cannabis, mulberry.

2. Dichogamy: Time of anther dehiscence and stigma receptivity are different forcing them for cross-pollination. The time gap between the two may vary from one day to many days. In protoandry anthers dehisce earlier than the stigma receptivity e.g; maize, sunflower. In protogyny stigma become recetive earlier than the anther dehisce e.g., Pearl millet mirabilis.

3. Self-incompatibility: Self fertilization in avoided by recognizing the self pollen by the stigma. E.g., Brassica, Petunia, Lilium.

4. Herkogamy: There is spatial separation of the anthers and stigma. Their relative position is such that self fertilization cannot occur. The stigma projects beyond the anthers and therefore pollen cannot land on stigma. E.g., Lucerne stigma is covered with a waxy film. The stigma does not become receptive until this waxy membrane is broken by visit of honeybees resulting in cross-pollination.

5. Male sterility: Absence or atropy or mis or malformed of male sex organ (functional pollen) in normal bisexual flower. Male sterility is of three types: genetic male sterility, cytoplasm sterility and cytoplasmic- genetic male sterility.

6. A combination of two or more of the above mechanisms may occur in some species. This improves the efficiency of the system in promoting cross-pollination.

Among different techniques self-incompatibility and sex expression have significance particularly in vegetable and flower hybrid seed production.

Hand Emasculation and Pollination

Hybrid seeds are produced manually by modifying the plant structure by removal of male organ from female plant before anthesis. This system is possible only when the male and female parts of a single flower or plants are separate. This is being adopted in bisexual perfect flowers where the androecium is removal with case. By removing the anther column / or male part from female line, the sterility of female line is created and is dusted with the pollen of desired male parent.

Self Incompatibility

Self-incompatibility is a mechanism, which avoids self fertilization through recognition of self pollen in or on stigma on the female pistil. But when pollen from other plant carried by wind or insects are accepted and sets seeds.

Self-incompatibility will prevents self pollination (inbreeding) and promotes crosspollination (out breeding) and creates genetic variability. SI is seen in hermaphrodite and homomorphic flowers. Self-incompatibility is a widespread mechanism in flowering plants that prevents inbreeding and promotes outcrossing. The self-incompatibility response is genetically controlled by one or more multi-allelic loci, and relies on a series of complex cellular interactions between the self-incompatible pollen and pistil.

Types of Self-Incompatibility (SI)

Heteromorphic Self-incompatibility

In this system flowers are of different morphology of the reproductive parts. The morphological differences can be seen visibly in flowers this will coincide with crossibility. The characters affecting this type of SI are style length, filament length, pollen size, exine sculpturing. The presence or absence of other SI mechanisms will not affect cross pollination in heteromorphic SI.

- Heteromorphic distyle: e.g., Primula

The flowers of primula have style at two different heights. There are two types of flowers.

1. Thrum flower: short style and long anther.

2. Pin flower: has short anthers and long style.

But these two are cross compatible.

a. Pin x Pin (ss x ss): Incompatible mating.

b. Pin x Thrum (ss x Ss): Compatible mating.

c. Thrum x Pin (Ss x ss): Compatible mating.

d. Thrum x Thrum (Ss x Ss): Incompatible mating.

- Heteromorphic Tristyle: e.g., Lythrum

Three types of flowers:

a. Short style and stamens are mid and long.

b. Mid style and stamen are short and long.

c. Long style and stamen are short and mid.

When homomorphic flowers are crossed it results in incompatible mating and when hetromorphic flowers are crossed it results in compatible mating.

Homomorphic Flowers

Flowers morphological are same, so mating types cannot be recognized by morphological features. This types of self-incompatibility is controlled by alleles. For crossing parental alleles should be different, then only fertilization takes places and seed sets.

Two types of self-incompatibility:

1. Sporophytic self-incompatibility.

2. Gametophytic self-incompatibility.

Sporophytic Self-incompatibility (SSI)

SSI is less common or rare when compared to GSI. The rejection of pollen is controlled by S loci which is dominant. The dominance relationship are like $S_1 > S_2 > S_3 >$. This type of self-incompatibility is controlled by diploid genotypes of the sporophyte (pollen). Pollen will not germinate on the stigma of the flower that contains either of the two alleles in the pollen so, rejected.

Gametophytic Self-incompatibility (GSI)

GSSI is more common when compared to SSI. The rejection and acceptance of pollen is controlled by loci. Unlike SSI, incompatibility is controlled by haploid genotype of pollen itself.

S_1 pollen can germinate on pistil of S_1S_2 but due to common S_1 allele the pollen tube growth seizes. Similarly in S_2 the pollen tube growth seizes. When S_3 pollen come in contact with pistil of S_1S_2 there will be normal growth of pollen tube and fertilization takes place, this is called as partial compatibility. Where as, S_3 and S_4 pollen can pierce in to the style ($S_1 S_2$) and cause fertilization, called as complete compatibility.

In single gene system there are three types of mating systems they are:

- S_1S_2 X S_1S_2: 0% compatibility.

- S_1S_2 X S_2S_3: 50% compatibility.

- S_1S_2 X S_3S_4: 100% compatibility.

Pollen rejection and acceptance of cross pollen is a mechanism of complex interactions.

There are three main interactions:

1. Pollen stigma interaction seen in SSI.

2. Pollen- style interaction seen in GSI.

3. Pollen tube- ovule interaction seen in GSI.

Pollen tube- ovule interaction is very rare and seen in some cases of cocoa, pollen tube reaches ovule and effect fertilization but incompatible combination will degenerate the embryo at early stage development only resulting in no seed set.

Pollen-stigma Interaction

This type of interaction is seen in SSI. It includes dry stigma. The stigma has a hydrated layer of proteins know as pellicle. This pellicle is involved in incompatible reaction. Within a few minutes of pollen reaching stigmatic surface, the pollen releases an exine exudates which is glycoprotein. Due to allele proteins interactions induces immediate callose formation in the papillae, this is in direct contact with pollen. Also this callose is deposited on the protruding pollen tube preventing further germination of the pollen. Here stigma is the site of incompatibility reaction. If once pollen cross this stigmatic barrier, there is no further rejection of pollen.

In case of wet stigma seen in GSI, when pollen sits on the stigma, pollen coat stimulate an unusual form of stigmatic reaction focused beneath the areas of coating. This results in the deposition of fibrogranular electron opaque layer by the extracellular vesicles in the outer layer of the papillae wall. This results in blockage of water passage or loss of osmotic competence by pollen resulting in drying of pollen.

Pollen Tube - style Interaction

This is common in gametophytic self-incompatibility. The pollen can germinate and pollen tube growth will pierce the stigma. But the rate of the pollen tube growth is very slow when compared to compatible pollen. This slow growing pollen tube will stops due to exhaustion of reserve materials and deposition of allele polysaccharide at the tip of the pollen tube which blocks the growth of tube. This deposition is limited to the tip and thus will not affect the other compatible pollen tube growth.

Three-way Hybrid

Two SI lines are crossed to get an heterozygote and again crossed with self-compatible parent which has suppressors locus and give F1 hybrid which is self-compatible.

Triline Hybrid

Hybrid seed production is done by crossing of two inbreed lines. Have to maintenance these inbreed lines which are self-incompatible. For success of self pollination need to eliminate SI. There are many methods to over come SI

- Bud pollination.

- Washing stigma surface/ pollen grains with organic solvents.

- Mechanical and electric methods.

- CO_2 Treatment.

- High temperature treatment.

Modification of Sex

Hermophrodite flowers has both male and female reproductive organ in a single flower.

1. Complete flower; flowers containing all the four whorls viz., sepals, petals, androecium and gyneocium.

2. Incomplete flower: flowers missing any of the one whorl.

3. Imperfect Flowers- Sexual distinctness: Monoecious (Maize) vs. Dioecious (Hollies, Poplars)

4. Perfect Flowers- Gamete Maturation Time (Dicogamy).

Production of this unisexual/imperfect flowers will naturally leads:

- Outcrossing avoids the deleterious effects of inbreeding depression,

- Promotes heterozygosity, genetic variability, and genetic exchange,

- Advantageous to the long-term survival and adaptation of a species.

Monoecious

Flowers are unisexual and are present at different position on the same plant. E.g. cucumber. Terminal flowers are male flower. In the middle of the plant is female favouring crosspollination.

Dioecious: male flowers and female flowers are in different plant. So called as male plant and female plant.

Sex Modification Through Hormones and Chemicals

Sex expression in dioecious and monoecious plants is genetically determined and can be modified to a considerable extent by environmental and introduced factors such as mineral nutrition, photoperiod, temperature, phytohormones. Amongst these, phytohormones have been found to be most effective agents for sex modification and their role in regulation of sex expression in flowering plants has been documented. The morphological differences in various sex types and their specific metabolic characteristics result from the possession of specific patterns of proteins, enzymes and other molecules. Modification of sex expression in cucurbits has been induced both by changing the environmental conditions and by applying treatments with growth regulators. Auxin treatments increase the female sex tendency while gibberellins cause a shift towards maleness.

- Hormones & Chemicals inducing Femaleness: Auxins- NAA, Etherl, Ethephon, Cytokinis- BA, Brassinosteriods etc.

- Hormones & Chemicals inducing Maleness: GA3, AgNO3, ABA Thio porpinic acid, Pthalimide, Paclobutrazol etc.

In cucumber AgNO3 found to be potent inhibitors of ethylene action leading to femaleness. It should be sprayed when first true leaf is fully expanded. Gibberlic acid spray will leads to excessive elongation and weakening of plants and there will be increased number of mall formed male flowers with less pollen. In gynoecious cucumber there will be increased number of male nodes when sprayed with silvernitrate and gibberlic acid, which made possible for multiplication of gynoecious in hybrid seed production.

Environmental Sex Modification

Environment has greater influence on the sex modification. But due to introduction of photosensitive varieties or hybrids in modern era of agriculture it has gained less importance. However in seed production it has its influence on sex expression. In cucumber, high temperature and long day length (> 14 hours) favors male flowers. High temperature will extends the flowering of female flowers. As the temperature increases form 19° C to 23° C the node for first female flowers has also increased from 9.6 to 16.5 number. This clearly indicates that high temperature favors male and delays female flowering.

Male sex expression of several plant species is favored by high temperatures and female sex expression by low temperatures. Male sterile mutant of tomato developing male sterile flowers at a minimum temperature of 30°C and normal flowers at lower temperatures. In Brussels sprouts of low temperature effect on the development of the androecium. In onions a slight production of viable pollen by normally male sterile plants above 20 °C.

Male Sterility

Hybrid production requires a female plant in which no viable male gametes are borne. Emasculation is done to make a plant devoid of pollen so that it is made female. Another simple way to establish a female line for hybrid seed production is to identify or create a line that is unable to produce viable pollen. This male sterile line is therefore unable to self-pollinate and seed formation is dependent upon pollen from the male line.

In hermaphrodite flowers pollens are non-functional or inactive or sterile while, female gametes functions normally. It is the inability of plant to produce or to release functional pollen as a result of failure of formation or development of functional stamens, microspores or gametes. Male sterility can be either genetic or cytoplasmic or cytoplasmic-genetic. This prevents autogamy and permits crosspollination. Promotes heterozygosity. Sterility is due to nuclear genes or Cytoplasmic gene or both.

In hybrid seed production process female is a male sterile line crossed with male fertility restorer line to get hybrid.

Cytoplasmic Male Sterility

Cytoplasmic male sterility, as the name indicates, is under extra nuclear genetic control mainly mitochondrial genome. They show non-mendelian inheritance and are under the regulation of cytoplasmic factors. In this type, male sterility is inherited maternally. In general there are two types of cytoplasm: N (normal) and the aberrant S (sterile) cytoplasms. These types exhibit reciprocal differences. Cytoplasmic male sterility (CMS) is caused by the extra nuclear genome (mitochondria or chloroplast) and shows maternal inheritance. Manifestation of male sterility in CMS may be either entirely controlled by cytoplasmic factors or by the interaction between cytoplasmic and nuclear factors.

- Stamen (anther and filament) and pollen grains are affected.

- It is divided into:

  ○ Autoplasmic.

CMS has arisen within a species as a result of spontaneous mutational changes in the cytoplasm, most likely in the mitochondrial genome

- Alloplasmic.

CMS has arisen from intergeneric, interpecific or occasionally intraspecific crosses and where the male sterility can be interpreted as being due to incompatibility or poor co-operation between nuclear genome of one species and the organellar genome another CMS can be a result of interspecific protoplast fusion. Cytoplasmic male sterility is used in hybrid seed production. In this case, the sterility is transmitted only through the female and all progeny will be sterile. This is not a problem for crops such as onions or carrots where the commodity harvested from the F1 generation is produced during vegetative growth. These CMS lines must be maintained by repeated crossing to a sister line (known as the maintainer line) that is genetically identical except that it possesses normal cytoplasm and is therefore male fertile.

Disadvantages

1. Insufficient or unstable male sterile.

2. Difficulties in restoration system.

3. Difficulties with seed production.

Cytoplasmic-genetic Male Sterility

Male sterility is controlled by an extranuclear genome and often nuclear genes can have the capability to restore fertility. When nuclear restorations of fertility genes are available for CMS system in any crop, it is cytoplasmic-genetic male sterility; the sterility is manifested by the influence of both nuclear (Mendelian inheritance) and cytoplasmic (maternally inherited) genes. There are also restorers of fertility (Rf) genes, which are distinct from genetic male sterility genes. The Rf genes do not have any expression of their own unless the sterile cytoplasm is present. Rf genes are required to restore fertility in S cytoplasm, which causes sterility. Thus N cytoplasm is always fertile and S cytoplasm with genotype Rf- produces fertile; while S cytoplasm with rfrf produces only male steriles. Another feature of these systems is that Rf mutations (i.e., mutations to rf or no fertility restoration) are frequent, so N cytoplasm with Rfrf is best for stable fertility.

Cytoplasmic-genetic male sterility systems are widely exploited in crop plants for hybrid breeding due to the convenience to control the sterility expression by manipulating the gene cytoplasm combinations in any selected genotype. Incorporation of these systems for male sterility evades the need for emasculation in cross-pollinated species, thus encouraging cross breeding producing only hybrid seeds under natural conditions.

In cytoplasmic-genetic male sterility restoration of fertility is done using restorer lines carrying nuclear restorer genes in crops. The male sterile line is maintained by crossing with a maintainer line which has the same genome as that of the MS line but carrying normal fertile cytoplasm.

Genetic Male Sterility

Male sterility is controlled by mutations in nuclear genes in the single recessive genes affect stamen and pollen development, but it can be regulated also by dominant genes. MS alleles are generally recessive. A male sterile line is maintained by crossing with heterozygous male fertile line.

Male sterile plants of monoecious or hermaprodite crops are potentially useful in hybrid program because they eliminate the labor intensive process of flower emasculation Constraint of the use of genetic male sterility.

- The maintenance of the male sterile line. Normally, a GMS line (A-line) is maintained by backcrossing with the heterozygote B-lines (Maintainer lines), but the progeny produced are 50% fertile and 50% male sterile.

Breeding of Clonally Propagated Plants

The definition of a clonally propagated crop is that the material to cultivate and maintain a variety is obtained by asexual reproduction, regardless of how different the plant material used for propagation is within and between species, encompassing tubers, roots, stem cuttings and corms, as well as asexually developed seeds (seeds developed without meiosis). It should be remembered that if crops such as maize (bred as an open-pollinated or hybrid crop) or beans (bred as a cross-fertilized, self-fertilized or hybrid crop) were to be propagated by stem cuttings or asexually developed seeds, they would be clonally propagated crops. In contrast, in breeding clonally propagated crops, the breeding techniques and methods that are usually associated with cross-fertilized and hybrid crops can be very useful. An example is the selection of parents in potato, cassava and sweet potato breeding, which are recombined in open-pollinated polycross nurseries to create new genetic variation. It is almost certain that techniques and methods from breeding cross-fertilized and hybrid crops will become much more important in the future of clone breeding.

It appears to be simple: to break the normal clonal propagation by a crossing step, and thus develop sexual seeds and genetic variation from which to select new clones. All propagation steps from the first to the last selection step are again 'normal' asexual reproduction. Hence, the finally selected clone is genetically identical with the original seed plant from which the selected clone is derived. In other words, each seed plant is a potential variety. Roots and tubers, fruit and tree plant species have been used by human since long before the dawn of agriculture. They have been domesticated by IPB and several made a substantial yield progress by FPB in some regions of the world. However, in other regions of the world there is not much yield progress, and in these regions there appears to be a clear need of PPB for progress.

An example of the needs and requirements of clonally propagated varieties can be found in potato (Solanum spp.). There are about 200 wild potato species. They usually contain glycoalkaloids, which give tubers a bitter taste and which are toxic when consumed in large quantities. It is nearly certain that 100 to 130 centuries ago indigenous knowledge in the Andes and along the Pacific coast of South America was those sites where it was possible to collect wild potato tubers where species and mutants were growing that had low alkaloid content. Although these tubers were very small, the man, or more probably a woman, made life much easier by growing and maintaining desirable types by cloning close to their homes. This happened more than 8 000 years ago, and most likely independently at several places. Those types were preferred that were easier to maintain, easier to harvest (shorter stolons) and had larger tubers compared to other types. The result was the domestication of pitiquiña (Solanum stenotomum), which was most probably selected from S. leptophyes or S. canasense. From the view-point of the knowledge of the twenty-first century it is not surprising that suddenly potato plants with larger leaves and larger tubers were found. Potato spontaneously changes its polyploidy level by unreduced gametes and recombination. Polyploid potatoes are more vigorous than their diploid ancestors. The result was the domesticated of polyploid andigena (S. tuberosum subsp. andigena). Andigena is the ancestor of the commercial potato in long-day temperate climates—the so-called Irish potato (S. tuberosum subsp. tuberosum). This IPB of potato and introductions of FV of potato into the Northern Hemisphere changed the world both socio-economically and politically.

References

- Breeding-self-pollinated-species, plant-breeding: britannica.com, Retrieved 31 March 2018

- Procedure-for-pureline-selection, pure-line-selection, plant-breeding: theagricos.com, Retrieved 15 April 2018

- Bulk-method-merits-applications, plant-breeding: theagricos.com, Retrieved 25 April 2018

- Hybridization-method-of-crop-improvement-17701: biologydiscussion.com, Retrieved 15 March 2018

- Top-3-methods-used-in-self-pollinated-crops-breeding-methods-60846: biologydiscussion.com, Retrieved 19 May 2018

- Top-3-Breeding-Methods-Used-for-Cross-Pollinated-Crops-367364157: scribd.com, Retrieved 18 July 2018

Permissions

All chapters in this book are published with permission under the Creative Commons Attribution Share Alike License or equivalent. Every chapter published in this book has been scrutinized by our experts. Their significance has been extensively debated. The topics covered herein carry significant information for a comprehensive understanding. They may even be implemented as practical applications or may be referred to as a beginning point for further studies.

We would like to thank the editorial team for lending their expertise to make the book truly unique. They have played a crucial role in the development of this book. Without their invaluable contributions this book wouldn't have been possible. They have made vital efforts to compile up to date information on the varied aspects of this subject to make this book a valuable addition to the collection of many professionals and students.

This book was conceptualized with the vision of imparting up-to-date and integrated information in this field. To ensure the same, a matchless editorial board was set up. Every individual on the board went through rigorous rounds of assessment to prove their worth. After which they invested a large part of their time researching and compiling the most relevant data for our readers.

The editorial board has been involved in producing this book since its inception. They have spent rigorous hours researching and exploring the diverse topics which have resulted in the successful publishing of this book. They have passed on their knowledge of decades through this book. To expedite this challenging task, the publisher supported the team at every step. A small team of assistant editors was also appointed to further simplify the editing procedure and attain best results for the readers.

Apart from the editorial board, the designing team has also invested a significant amount of their time in understanding the subject and creating the most relevant covers. They scrutinized every image to scout for the most suitable representation of the subject and create an appropriate cover for the book.

The publishing team has been an ardent support to the editorial, designing and production team. Their endless efforts to recruit the best for this project, has resulted in the accomplishment of this book. They are a veteran in the field of academics and their pool of knowledge is as vast as their experience in printing. Their expertise and guidance has proved useful at every step. Their uncompromising quality standards have made this book an exceptional effort. Their encouragement from time to time has been an inspiration for everyone.

The publisher and the editorial board hope that this book will prove to be a valuable piece of knowledge for students, practitioners and scholars across the globe.

Index

A

Abc Model, 55-57
Abscisic Acid, 16, 19, 42-43
Allelic Gene, 125, 127-128
Apical Cell, 15, 30-31
Apical Meristem, 16-18, 30-33, 43-44, 91
Apogamy, 76-77, 80, 97
Apomixis, 61, 63, 69, 75-79, 97
Apospory, 76, 78, 81, 97
Asexual Reproduction, 60-63, 67, 69-72, 76, 202
Auxin, 16, 18-19, 30, 33-34, 199

B

Bac Clone, 7
Back Cross Method, 153, 160
Basal Cell, 15, 31-32, 89-90
Budding, 61, 69, 77

C

Carotenoid, 20, 52, 54
Cell Division, 15, 17-21, 25, 29, 31, 34-35, 61, 64, 74, 99-100, 105-106, 137
Cell Growth, 18-20, 33
Chloroplast, 2, 48, 55, 106, 134-138, 200
Chromatid, 99-100, 117, 120
Chromosome Number, 2, 6, 98, 103, 107-108, 119-120
Class A Gene, 57-59
Colinearity, 5-7
Cross-pollination, 83-84, 96, 143-144, 167-169, 171, 189, 192, 194
Crossing Over, 110-111, 114-117
Cryptochrome, 46, 49, 52-53, 55
Cytokinin, 16, 19
Cytoplasmic Inheritance, 98, 122, 133, 135, 138

D

Differentiation, 14, 18-20, 25, 29, 34
Double Fertilization, 30, 66, 88-89, 91, 105

E

Emasculation, 140-144, 153, 194, 199, 201

Embryo Sac, 37-40, 66, 76-77, 81, 89
Embryogenesis, 14, 16, 25, 29-35, 59
Evolution, 1, 6, 21-22, 25, 27, 29, 42, 59, 119, 121, 132, 134, 144

F

Female Gametophyte, 37-38, 93
Flavin, 51, 53-54

G

Gametophyte, 34, 37-38, 48, 64, 76-77, 80, 93, 103
Gametophytic Apomixis, 76-78
Gene Interaction, 98, 124
Genetic Map, 4, 8
Genome Sequencing, 7-9
Gibberellin, 18-19, 79
Globular Stage, 30-32, 34
Grafting, 69, 71-72
Guard Cells, 20, 54-55

H

Heart Stage, 30-32
Heterochromatin, 43, 106-107
Heterosis, 2, 44, 166, 174, 178-179, 181-186, 188, 194
Histogenesis, 15, 18-19, 21
Homology, 6, 10, 28-29, 52, 59
Hybrid Varieties, 139, 146, 166, 182, 189
Hybridization, 1-2, 6-7, 11, 123, 139-140, 144, 146-147, 152-153, 156-157, 161, 165-166, 172, 182, 185, 203

I

Inbreeding, 146, 148-149, 163, 171-173, 176-177, 186-188, 195, 198

J

Juvenility, 25, 27, 59

L

Linkage, 109-113, 144, 148-149, 163, 179

M

Mass Selection, 146-150, 152, 181-183, 186

Megaspore, 37-39, 41, 77-78

Meiosis, 1, 36-37, 64, 77-78, 98-101, 103, 106, 110, 114-115, 117, 120, 163, 202

Meiosporogenesis, 35-36

Meristematic Tissue, 31-32, 44, 74

Monocot, 45, 90-92

Morphogenesis, 22, 25, 45-46, 48

Morphology, 14, 21-25, 28-30, 59, 120, 195

Mutation, 5, 57-59, 119-124, 132, 136, 138, 144, 147

N

Nuclear Division, 78, 103, 120

O

Organogenesis, 15, 17-19, 21, 25

P

Perennial, 61, 183, 187

Photo-morphogenesis, 45-46, 48

Photoreceptor, 46, 48-49, 51-53, 55

Phototropin, 52, 54

Phototropis, 54

Phytochrome, 46-50, 55

Plant Breeding, 1, 4, 10-12, 75, 83, 97, 123, 147, 150, 157, 172, 181, 185, 187

Plant Hormone, 19

Pleiotropy, 98, 132, 148-150

Polyploidy, 2-3, 42, 76, 107-108, 120, 203

Pure Line Selection, 150, 152

Q

Quantitative Character, 131

R

Recurrent Selection, 171, 173, 179-181, 187

Root Apical Meristem, 18, 31-32, 44

S

Seed Dormancy, 16, 42, 94

Seed Germination, 14, 16, 47-49, 73, 92

Seedling, 15-17, 27, 33, 45-47, 53, 66, 70, 89, 92, 124, 136

Self-pollination, 83-84, 140, 145, 168, 175-176

Sexual Reproduction, 60, 63-65, 69, 75-76, 79, 81-82, 96, 103, 108, 137

Shoot Apical Meristem, 16-17, 30-33, 43-44, 91

Single Seed Descent Method, 153, 158-159

Somatic Cell, 29, 121

Somatic Embryo, 30, 34

Somatic Embryogenesis, 29, 34

Sporogenesis, 14, 35-36, 62

Synthetic Variety, 172-173, 175, 182-184, 186-187

T

Transcription, 25, 41-43, 56-58, 104

Transgressive Segregation, 155, 164

V

Vascular Plant, 24

Vascular Tissue, 32, 90

Vegetative Propagation, 62-63, 67-69

Vegetative Reproduction, 60-61, 69

Z

Zeaxanthin, 52, 54-55